Success with Baby Chicks
by Robert Plamondon

Norton Creek Press
36475 Norton Creek Road
Blodgett, OR 97326

nortoncreekpress@plamondon.com
http://www.plamondon.com

Success With Baby Chicks

ISBN 0-9721770-0-0

2 4 6 8 10 9 7 5 3 1

Chapter 15. Health Problems and Predators   141

Chapter 16. Useful Tables   149

Chapter 17. Recommended Reading   150

Chapter 18. For More Information   155

Chapter 1. Introduction

Raising baby chicks can be a delightful experience if things go well, but it can be discouraging and even heartbreaking if things go wrong. The purpose of this book is to help you avoid the pitfalls and achieve the delightful experience that makes poultrykeeping attractive in the first place.

When baby chicks are brooded artificially—that is, when they are raised without their mothers—they depend on you for food, water, warmth, and protection. How well these are provided in the first few weeks of life will determine both their short-term survival and their long-term development.

When my wife Karen brought home our first group of twenty-five baby chicks, we had read in a book that chicks could be started out in a box heated with a sixty-watt light bulb. Nowhere did this book mention that this was adequate only if you brooded the chicks in a room heated to 70 °F! Our chicks were in a basement that maintained a steady 50-55 °F. The chicks weren't warm enough until we had escalated all the way to a 250-watt heat lamp. Fortunately, the chicks did very well in spite of our inexperience, giving us some much-needed but largely misplaced confidence.

A lot of our chick-rearing experiences were like this. We'd try to follow the best instructions available, only to find that these instructions were only a general outline, with important details missing. The result was that sometimes all our chicks would do well and sometimes they wouldn't.

At about the time we were becoming resigned to this, we met a woman who had retired from the egg business after raising 30,000 chicks per year for many years.

She told us proudly that she would lose about 1% of the new chicks during the first week, but that every chick that survived the first week would still be alive and healthy when they were moved into the laying house when they were five months old.

How was this possible? Our own results couldn't hold a candle to this. She explained that it was largely a matter of reducing things to a system. If something works, try to do it the same way next time. If something seems troublesome, try to find a better way, and always stay on the lookout for new ideas. She attended poultry science classes at the nearby campus of Oregon State University from time to time, and stayed in touch with the Poultry Specialists at the Extension Service

Her brooder houses and equipment weren't fancy at all. She used overhead infrared heat lamps to provide brooder heat, which is what most hobbyists and small flockowners use. In other words, this was not a question of fancy equipment, but of knowledge and skill.

I was impressed by what I heard, and rethought my own chick-brooding techniques. I went to the University library and read everything I could find, old and new, on brooding technique.

The results have been encouraging. For one thing, my reading dispelled the notion that I was a victim of a tall tale. While 1% mortality is good by anybody's standards, the mainstream egg industry considers 2% mortality to be achievable and 5% mortality to be pretty bad. On the other hand, many small flockowners would be happy if they could hold their average losses to 10%, and a great many people have given up on poultrykeeping altogether because they were discouraged at their inability to keep their chicks alive.

I also discovered that the methods that lead to success with baby chicks are neither difficult nor expensive. They used to be widely publicized, but the books describing them are no longer in print, so they have become the Lost Secrets of the Poultry Masters.

There was a golden age of poultrykeeping, running from roughly from 1900 to 1950. This was a period when most eggs and poultry still came from small commercial flocks on family farms, with an average flock size of under 100 hens through most of the period. It was also the early period of poultry science, when researchers at the Experiment Stations tried out every technique they could think of, to see which ones worked. With millions of farms raising chicks every year, there was a lot of business for equipment manufacturers, and competition was intense. There was a greater selection of books and equipment on the market during that period than at any time before or since.

With the decline of the family farm, whole categories of knowledge and equipment have vanished. You used to be able to buy insulated chick brooders for cold-weather brooding with either propane or electric heat. Now you can't buy insulated brooders at all. This is one reason why small-flock chick rearing is harder than it used to be.

Success with brooding depends on believing that it can be done. Once I believed it, I started looking with a keener eye at my setup and methods, and this was as important as any specific techniques.

Like a lot of people who have gotten serious about being successful with chicks, I have reached a point where I can get a typical batch of chicks through the brooding period without losses after the first few days, and even this first-week mortality is a lot lower than it used to be. The hatchery we use adds extra chicks to cover losses in shipping, and the extras frequently exceed the losses both in shipping and in brooding. As often as not, I discover that my batch of 75 chicks has 76 survivors at the end of the brooding period! Things don't work this well every time, but the progress since I started out has been tremendous.

The impact of these results is not easy to exaggerate. I used to dread brooding chicks because the results were so erratic. It's very discouraging when baby chicks didn't do well. Now that the results are consistently good (barring the occasional misfortune), I can brood chicks with enthusiasm.

My goal in this book is to set down both the general principles and the little details that can help you be succeed in brooding poultry, and to set down as many different techniques as I can, to make your chick-rearing as successful and as enjoyable as possible.

Chapter 2. Brooding Quick-Start

2.1. Before Ordering the Chicks

2.1.1. Prepare the Brooder House

1. If you don't already have a brooder house, make one or adapt an existing structure. See Chapter 14.
2. Clear away any brush or trash that may have accumulated around the brooder house.
3. Examine the brooder house for leaks in the roof, gaps in the floor, and rat holes—and fix them.
4. If there are signs of rats, trap or poison them now.
5. If there is an infestation of roost mites or other noxious bugs, treat the brooder house now. See Chapter 15.
6. If there is old litter in the house, decide whether you are going to re-use it. If so, prepare it as described in Chapter 13. Otherwise, remove the old litter and put in new.
7. Acquire a brooder, draft guard, first feeders, and first waterers. See Chapter 5.
8. Dump any feed left over from last time.
9. Close up the brooder house by closing all the windows and covering any sizable openings with tarps, sheets of plastic, or plastic feed sacks.

> **DANGER!** If you are using vent-free propane brooders, it is possible for carbon monoxide to build up to lethal levels in a tightly closed brooder house. Install a carbon monoxide alarm if you're going to use propane brooders in a tight house.

2.1.2. Management Decisions

1. Did you have high mortality, a wet house, caked litter, cannibalism, or coccidiosis last time? Consider brooding fewer chicks per house this time.
2. Plan out your brooding season.
3. Select a hatchery. See Chapter 3.

4. Consider trying a new breed by placing an order that consists of your favorite breed plus one that might be a contender, so you can raise them together and see which one you like best. Sometimes you can become a lot happier with poultrykeeping just by switching to a breed that suits you better.
5. Did your feeders or waterers frequently go empty? Did your waterers cause you trouble? Was there feed spillage? Are you happy with your brooder? Is there mud in front of your brooder house? Do you have a conveniently placed trash can? Does the water piping to the brooder house pass muster? Does the wiring to your brooder house make you nervous? Is your brooder house permeable to rodents, pets, or predators? Now's the time to make changes.
6. Is it time to build another brooder house or acquire some battery brooders?
7. Do you have enough housing for the chicks after they leave the brooder house?

2.2. Ordering the Chicks

1. Write down the expected date of arrival and be sure a phone call from the post office will reach you.
2. While you're at it, write down the type of chicks ordered and the hatchery they were ordered from. If you've been considering more than one breed or more than one hatchery, it's easy get lose track of what you really ordered.

2.3. Before the Chicks Arrive

1. If this is the first batch of the season, start the brooder several days before the chicks are due to arrive to make sure it still works. Otherwise, start it at least 24 hours in advance.
2. Buy feed. See Chapter 11.
3. Buy or make feeders and waterers. See Chapters 11 and 12.
4. Check the temperature under the brooder to make sure everything is okay. Do this enough in advance that you can do whatever it takes to keep from being chilled after they arrive.

5. The floor under the brooder must be warm and dry to the touch before the chicks arrive. This is very important.
6. Install a draft guard, 10-18 inches high, around the brooder, with 2-3 feet of space between the edge of the brooder and the draft guard.
7. Make sure there's plenty of light for the chicks to see by. They can't eat or drink in the dark.
8. Clean your quart-jar waterers and (if they are reusable) your first feeders. See Figure 1.
9. Clean, inspect, and repair your automatic watering system (if any), feed troughs, tube feeders, "practice perches," waterer stands, and other equipment that will be brought into use as the chicks get older.
10. Double-check that your brooder is set up for day-old chicks, and has not been left the way it was the last time it was used, throttled back for older chicks who barely needed any heat.

Figure 1. A brooder area ready for the chicks. Box lids are set out as temporary feeders. Quart-jar waterers on set up on little wooden frames covered with hardware cloth. (This photo, and others in the sequence are from Rice & Botsford's Practical Poultry Management, Sixth Ed.,1956, pp. 3-10)

2.4. When the Chicks Arrive

1. If you fetch the chicks from the Post Office, run the heater in your car to keep them warm on the drive home if the weather is cool. If it's warm, keep the chick box out of the sun.
2. Place the chicks under the brooder without delay. Don't leave the brooder house door open any more than absolutely necessary. Commercial chicken farmers simply turn the chick boxes upside down to dump the chicks under the brooders. This doesn't harm them, and gets them where it's warm with a minimum of delay.
3. Give the chicks warm water to drink immediately in quart-jar waterers, with at least one waterer for every 25 chicks. One waterer per 15 chicks is better. After chilling, dehydration is your biggest worry.
4. Give the chicks feed in the first feeders either immediately or after three hours (opinions vary). The 3-hour delay is intended to get the dehydration problem taken care of before the issue becomes complicated by feed. First feeders can be egg flats (1 for every 50 chicks), plastic cafeteria trays (1 for every 50 chicks), or the lid or bottom of the box the chicks arrived in.

2.5. Days 1-2

1. Don't let the chicks get chilled. Check on the chicks several times per day. Move any that get lost back into the heat. Make sure they are warm enough. Make a special trip at nightfall to make sure all the chicks make it back under the brooder. Also check first thing in the morning to make sure they're warm enough.
2. Spend time with the chicks. This is vitally important at all ages. If you deal with the chicks hurriedly or mechanically, all the fun goes out of poultrykeeping. Also, when things start to go wrong, you won't notice. Take a few extra minutes each time you're in the brooder house.
3. Leave the lights on all night so the chicks can see to eat and drink. It's not time to put them on a day/night cycle yet.

4. Refill the waterers and feeders as necessary. The chicks will kick feed out of the first feeders, and it will be lost. Don't try to prevent this.

5. Each time you visit the brooder house, check under the brooder for sick or dead chicks. Dead chicks need to be removed immediately. Some first-week mortality is normal. The amount of it depends on the amount of stress the chicks underwent during shipping and the amount of stress in the brooder house. The highest mortality will almost always be in the first 48 hours. It should cease, or almost cease, after that.

Figure 2. Day 1.

2.6. Days 3-4

1. Check on the chicks at least twice per day. Take your time.
2. Keep in mind that chickens are easily stressed by sudden changes in routine. Changes must be made gradually.
3. Expand the draft guard to give the chicks more space and to make room for more equipment.
4. Add larger feeders, either chick troughs or small (15 lb.) hanging tube feeders. Tube feeders should start with their feed pans flat on the ground. Troughs should be filled to the top. Use eight feet of chick trough per 100 chicks (two 4-foot

troughs, four two-foot troughs, or eight one-foot troughs), arranged so the chicks can feed from both sides, or two tube feeder for every 100 chicks. Keep using the first feeders.

5. Add larger waterers, either chick founts (which come in 1-, 3-, and 5-gallon sizes, 1 gallon per 50 chicks) or automatic waterers. These should be on stands that keep them above the floor and prevent litter from getting in the water. The waterers should be adjusted on the high side, so the chicks have to stretch a little to get the water. This will prevent them from splashing in it and getting chilled. See Chapter 12. Don't remove any of the quart-jar waterers yet.

6. Discontinue all-night lights after three nights.

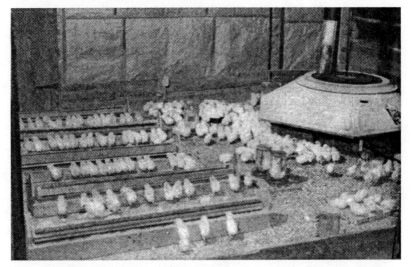

Figure 3. Day 3. The draft guard has been expanded and trough feeders have been installed (these days, tube feeders are more popular). Box feeders are still used.

2.7. Days 5-10

1. Expand the draft guard again on Day 5. If the chicks are getting past it, or the house is so small practically all of it is inside the draft guard already, remove it.

2. Remove the quart-jar waterers gradually, one or two per day, until only the large-capacity or automatic waterers remain.

Keep an eye on the chicks; sometimes it takes longer for them to use the big waterers, and you'll have to hold off removing the small ones.

3. Remove the first feeders gradually, one or two per day, until only the trough or tube feeders remain.
4. If tube feeders are used, check their height each day, adjusting them so the chicks are neither straining up nor reaching down to eat.
5. If trough feeders are used, fill them a little less full day by day, because a full trough leads to a great deal of feed wastage.
6. If overhead heat-lamp brooders are used, raise them a couple of inches higher at the end of the first week. If insulated heat lamp brooders are used, reduce the wattage of the bulbs at the end of the first week if the chicks seem comfortable. Turn down thermostatically controlled brooders by 5 °F.
7. By the end of the first week, mortality should have ceased altogether. If not see Chapter 15.

2.8. Day 11

1. Double the amount of feeder space. If using trough feeders, it may be time to replace them with ones designed for larger chicks. This will reduce feed wastage. Continue increasing the height of the feeders as the chicks grow.

2.9. Remainder of Second Week

1. Pay attention to litter quality. Caked litter tends to appear around the brooder at this time, and wet litter tends to appear around the waterers. Remove both as they appear. See Chapter 13.
2. Start increasing ventilation a little at a time.
3. At the end of the second week (Day 14), turn down the thermostat another five degrees, raise overhead infrared heaters two inches, or raise insulated heat-lamp brooders an inch or two—whichever is appropriate to your brooder.

Figure 4. The same chicks during week 2. They have been given more space and larger waterers on wire stands.

2.10. Third Week

1. Except for broilers, add some "practice perches" to encourage early roosting.
2. Turn the thermostat down or raise the brooder again, as appropriate.
3. Increase ventilation some more.
4. Keep checking the brooder house twice daily. It's easy to fall out of the habit because this period is generally trouble-free.

2.11. Weeks 4-5

These are the last weeks of the brooding period. Depending on the weather, broilers may not need brooder heat after two weeks, Leghorns after three, and other breeds after four. But be prepared to give brooder heat to broilers for three weeks and other breeds for five. Add an additional week if you are brooding in winter.

Figure 5. Chicks sleeping at night. Their heat needs have gone down, and the space under the brooder has become too warm for comfort, so they sleep just outside.

The birds get quite large during this period, and a brooder house that was fine yesterday can be crowded today. Crowding can lead to sudden outbreaks of coccidiosis (a protozoan infection), feather-picking, and even cannibalism. It is very important to have enough floor space to keep the birds happy and healthy for the entire brooding period. This is easy if they are being brooded in the same house in which they will live throughout their lives, but if they are to be moved to different quarters, they need to be moved on time. Delay can be disastrous.

At the end of the brooding period, the feeders need to be swapped for larger ones that are suitable for adult birds, and (except for broilers) full-sized perches installed. For hens, nest boxes will be needed by Week 18 for commercial layers, or Week 20 for other breeds.

Figure 6. As the chicks grow, they need a lot more space. Note that the original troughs have been replaced with larger ones.

Figure 7. As these pullets approach laying age, they are given still larger feed troughs, this time set up on stands to make them easier to fill. That makes three sets of troughs, two of which are used during the brooding period.

Chapter 3. Breed and Hatchery Selection

"What kind of chicken should I buy?"

Well, it all depends.

People who farm for profit must select breeds carefully, as most breeds are low-producing and have virtually no profit potential. This question is difficult enough that I've given it a chapter to itself—Chapter 4.

People who are keeping poultry for pleasure can choose whichever breed they like. In this chapter, I'll try to help you find a breed that matches your goals. However, a lot depends simply on what strikes your fancy, and I can't give you any guidance there.

3.1. Breeds and Strains

Chickens are divided into *breeds* and *strains*. Basically, breeds are based on what the chickens look like, while strains are distinct family lines within the breed. The American Poultry Association's *Standard of Perfection* defines the universally accepted breeds, but there are "unofficial" breeds as well.

Unlike many other kinds of standard-bred livestock, a chicken belongs to whatever breed it appears to belong to, regardless of its bloodline. Pedigrees are unnecessary, and in fact inventive cross-breeding is frequently used by poultry fanciers to try to create show-winning birds. This means that birds from the same breed can have very different pedigrees, personalities, and productivity.

Strains are generally synonymous with the breeder producing the stock. For example, a "Rowley New Hampshire" is a New Hampshire Red descended from Mr. Rowley's flock. Because every bird in a strain is related, strains are usually much more uniform than breeds in their characteristics.

3.1.1. Important Strain Characteristics

There following five traits that I consider to be of particular importance:

1. *High-production.* A commercial egg flock needs to produce as many eggs as possible and commercial broiler flocks need to grow as quickly as possible, or they won't be profitable. On the other hand, if you're a hobbyist, backyarder, or fancier, high production may be more of a burden than a blessing. There are only so many eggs you can eat. In this case, choosing a low-producing strain can simplify your life.

2. *Non-cannibalistic.* Some strains of chickens are intensely cannibalistic and pluck each others' feathers out or even kill each other from an early age. This tendency has been successfully bred out of broilers, but the mainstream poultry industry uses beak-trimming more or less universally with its layers. Removing the top of the beak turns it from a sharp to a blunt instrument and greatly reduces the damage chickens can do to one another. But beak-trimming, in addition to being unaesthetic, is a very disagreeable chore and requires that you invest in an electric beak-trimming device. If you use cannibalistic strains, beak trimming is probably necessary. However, many strains of poultry are non-cannibalistic, and will behave themselves without beak-trimming. I think it's best to confine your choices to these breeds, at least to start with.

3. *Docile.* I wear gloves when I collect eggs, because some of my hens peck savagely at my hands when I try to get the eggs out from under them. In some strains this is common, while in others it is rare. Similarly, some breeds are tame, while others are panicky and wild. It's a lot more pleasant to have an egg flock when the hens are docile and friendly. This is not generally a problem with meat birds. Birds I have found docile and pleasant to have around include Barred Rocks, Wyandottes, Orpingtons, Australorps, and Jersey Giants. Some hybrid layers are also quite docile (but others aren't).

4. *Non-flighty.* If your birds fly easily, they can make your life miserable, especially if they're kept within about a hundred yards of your house. Leghorns fly easily, and go right over fences

15

that will hold in other breeds. They also fly very early. I have had two-week-old Leghorns fly right out of my brooder house, never to be seen again. Bantams also tend to fly well.

5. *Sex-linked.* A minor point is whether the sex of the birds can be determined easily right after hatching. In birds that can be color-sexed, the two sexes have very different-looking plumage right after hatching. This can be important when you buy sexed pullets, because the hatchery makes fewer sexing errors. When I buy 100 sexed Black Sex-Link pullets, I get no more than one cockerel. When I buy 100 sexed Production Reds, I get around five cockerels. Not only do I lose the production of the pullets I paid for but didn't receive, but I have to get rid of the cockerels I didn't want. This is also of importance to people with backyard flocks in towns with ordinances against roosters. Not everyone is willing to butcher the cockerels they end up with by mistake.

Because many of these important characteristics vary from strain to strain, the only safe thing to do is to call up the hatchery and ask them some questions about the breeds you're interested in. If the person at the hatchery doesn't know the answer to the question, ask to have the manager call you back. Sometimes excellent hatcheries have unskilled people answering the phone, but you shouldn't let this discourage you.

3.2. Sexed Chicks or Straight-Run?

If you are primarily interested in eggs, you should buy female chicks (pullets). If you buy straight-run (mixed sex) chicks, you need twice as much brooder space and twice as much equipment for the same number of eggs. Even with the extra equipment, your results will tend to be inferior to what they'd be if you bought pullets only, since larger chicks bully the smaller ones, and all the largest chicks are going to be the males (cockerels).

Some people want to order at least a few cockerels so they can breed from their flock, or so they'll have fertile eggs. My experience is that you can trust the hatchery to mis-sex enough chicks that this will not be a problem. If cockerels make up 5% of your flock, hatchability will be pretty high. It's surprisingly high even at 1%.

If you are more interested in meat than eggs, you should probably order broiler chicks rather than cockerels of the standard breeds. Modern broilers grow much more quickly than the old breeds, and will reach slaughter age in 6-8 weeks. Standard-breed cockerels are more personable than broilers, but they're only about half the size when they reach slaughter age at 8-12 weeks, so you have to do twice as much butchering for the same amount of meat.

By the way, if you want to order sexed chicks from a variety that's listed as being "straight run only," call up the hatchery and ask for pricing on sexed chicks as a special order. I have had good luck getting sexed pullets regardless of what the catalog said.

3.3. How Many?

If you are buying chicks by mail, they need to be ordered in increments of 25 chicks. There need to be enough chicks in the box to keep themselves warm by huddling together, but not so many that some of them get suffocated (chickens have weak lungs). The optimum number is 25, and most hatcheries won't even consider selling chicks except in increments of 25.

The shipping boxes for chicks come in 25, 50, and 100-chick sizes. The larger boxes are divided into 25-chick compartments. The price per chick is lowest if you buy 100 at a time.

If you're buying chicks locally, such as at the feed store, you can buy as few or as many as you like.

3.3.1. How Many Hens?

Figure 1 gives a pretty good ballpark figure for the number of hens you need for egg production.

Egg production is seasonal, with high production in the spring and summer and low production in the fall and winter. The number of hens in the table are enough to produce twice the desired total, in an attempt to keep the fall and winter shortages down to a

Doz. Eggs Eaten per Week	Modern Hybrid Layers	Old-Style Commercial Breeds*	All Other Breeds
1	5	7	10
2	10	14	20
3	15	21	30
* New Hampshire Reds, Rhode Island Reds, Standard Leghorns, California Grays, Barred Rocks, White Rocks, Anconas, White Wyandottes, and Delawares.			

Figure 1. Number of hens needed to support a given level of egg consumption

bearable level. I am assuming 240 eggs per year for modern hybrid layers, 180 for old-style commercial breeds, and 120 for everything else. Modern hybrid layers often lay in excess of 300 eggs per year in controlled-environment housing, but they will lay fewer in typical small-flock accommodations.

3.3.2. How Many Roosters?

If I were planning on incubating a lot of eggs, I would buy one cockerel for every five pullets, and butcher or otherwise dispose of the less-promising ones. I'd keep one cockerel for every ten hens or so for breeding.

3.3.3. How Many Broilers?

If you eat one chicken per week, a year's supply of broilers is about fifty birds. This assumes you have adequate freezer space and the desire to do that much butchering. Scale up or down from there according to your appetite and ambitions.

Poultry butchering is a skill that's acquired over time, and I would advise that you butcher as few birds as possible on your first try unless you have help from an expert. I've heard stories about people with no prior experience who tried to butcher fifty or more birds in a single day. One or two birds would be a better starting point.

You don't have to butcher a batch all at once. You can butcher a few at a time until they're all gone.

3.4. Buying Chicks at the Feed Store

In my area, the feed stores generally stock chicks only in March. Other parts of the country feature chicks for a longer season. The feed stores in my area can be counted upon to stock Cornish cross broiler chicks and two dual-purpose breeds with good egg-laying potential: Barred Rocks, and Production Reds.

3.4.1. Telling Pullets From Cockerels

Interestingly, both Barred Rocks and Production Reds can be sexed with good accuracy by anyone who knows the secret.

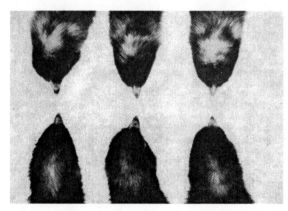

Figure 2. Sex-linked differences in the heads of Barred Rocks. Top, males; bottom, females. (Hutt, Genetics of the Fowl, p. 205)

Barred Rocks (and Black Sex-Links): The females have dark legs and relatively small, narrow white spot on the head. On Black Sex-Links the head spot may be missing altogether. The males have light legs and a large, splotchy white spot on the head. Also, the leg coloring on the females transitions sharply from black to yellow at the toes, while there is no sharp demarcation in the males. By looking at these characteristics, you ought to be able to distinguish the sexes with very few errors.

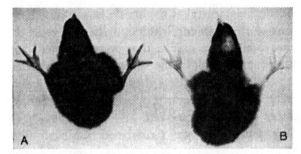

Figure 3. Sex-linked differences in the legs of Barred Rocks and Black Sex-Links. A, females; B, males. (MacIlraith & Petit, Ontario Agricultural College, 1940.)

Production Reds (also Rhode Island and New Hampshire Reds): Over 80% of the chicks with "chipmunk stripes" or small black spots on the back of the head are females. However, many female chicks have neither. Between 80% and 95% of the males (depending on breed) have a spot of white or creamy down on the wings. This may be large or small.[*]

3.5. Hatchery Selection

You should choose a hatchery based on location and reputation. Price is much less important.

For example, I read about an interesting experiment in an old poultry magazine. The author ordered 100 pullet chicks each from a number of hatcheries, which varied widely in price. The results are summarized in Figure 4.

Why was this true? Because discerning customers will pay top dollar for a top products, while cheapskates insist on paying low prices even if it means getting junk. A hatchery that loses the business of the discerning customers has the option of lowering its prices (and its quality) to satisfy the cheapskate market.

So should you buy from the most expensive hatchery you can find? You could do a lot worse.

---

[*] The information on sexing came from Hutt's *Genetics of the Fowl*, McGraw-Hill, 1949, pp. 196-199 and 205-206.

Category	Low-Priced Hatchery	High-Priced Hatchery
Number of extra chicks included	Few	Many
Chicks dead on arrival	Many	Few
Cockerels in an all-pullet order	Many	Few
Deaths during growing period	Many	Few
Egg production	Mediocre	Superior

Figure 4. What you can expect from low-priced and high-priced hatcheries

Perhaps the best way is to ask people who have kept chickens for several years which hatcheries they use, and what they think of them. Usually you don't have to ask very many people before a consensus emerges. Such a consensus is rarely wrong.

There are basically two kinds of commercial hatcheries: retail hatcheries that deal with backyarders, hobbyists, and small farmers, and mainstream commercial hatcheries that serve the giant commercial flocks. The chicks from the mainstream commercial hatcheries are generally very good and quite inexpensive, but these hatcheries are often unprepared to deal smoothly with small purchases (less than a truckload!), and you may have to bring your own box to carry them home in. The breed selection will be slim to none. Whatever they have will be a modern commercial breed of some sort. However, if such a hatchery is nearby and carries a breed that you like, this is probably for anyone with an egg or broiler flock who buys hundreds or thousands of chicks a year. Such hatcheries will hatch year-round.

Retail hatcheries are prepared to send chicks by mail, with a minimum order of 25 chicks. Most of them carry a wide variety of breeds, including several different commercial egg breeds. Some hatch commercial varieties year-round, though standard breeds are generally available only from late winter through the summer.

3.6. Receiving Chicks Through the Mail

As already mentioned, chicks are sent through the mail in specially designed shipping boxes for 25, 50, or 100 chicks. Hatcheries almost always have volume discounts, making it tempting to buy at least a hundred chicks at a time. Some hatcheries will give you wholesale pricing (the same prices they charge the feed store) once you have bought enough chicks. If you are buying hundreds of chicks a year, it won't hurt to ask.

My experience is that it's best to pick up chicks directly at the hatchery if the drive isn't more than an hour or two. The chicks seem stronger than ones that have been mailed. Chicks that are shipped by surface mail seem to arrive on my farm in better condition than those that have gone by air mail.

However, most of the chicks I have ever bought have come by air mail, and most of them have done just fine. Given the choice of two equally good hatcheries, one nearby, and one distant, I'd choose the nearby one. But if the distant hatchery seemed better to me, I'd buy from it without hesitation.

Sometimes a shipment is delayed in transit, and this can result in heavy mortality. I once received a shipment that was delayed an extra day, and it had 30% mortality in the first week. (The survivors developed normally). When the shipment arrives on time, however, the usual result is to lose 1% to 3% of the chicks in shipping and during the first week. Hatcheries add extra chicks, typically 4%, to cover normal losses. When I incubate eggs at home, first-week mortality is higher than this. My conclusion is that the hatchery's superior incubation is more than enough to compensate for the stress of shipping.

When you receive chicks through the mail, you will receive a phone call from the post office within minutes of the chicks' arrival, begging you to come pick them up. The noise of the chicks all peeping at once is cute for about ten seconds, then gets on everyone's nerves. When you order the chicks, make sure you give the hatchery a phone number that will actually reach you.

You may get a call from a main office rather than your branch post office. Our little Blodgett post office receives mail once a day, early in the morning, and if the chicks arrive at the regional hub too late for this delivery, I will get a call offering to let me pick up the chicks at the Corvallis post office (16 miles away) in the afternoon, rather than waiting for the chicks to arrive in Blodgett the next day. All in all, I've found the post office very obliging in fowarding my chicks by any means possible.

Chapter 4. Choosing Profitable Breeds

4.1. Introduction

Making money with poultry isn't easy. In the old days, just about every farm had chickens because they were a reliable money-maker. But the profit drained away from the old-fashioned farm flock in the 1950s and 1960s, and once the profit vanished, the flocks did, too.

The problem was that factory-farming produced eggs more cheaply and got them to market fresher than eggs that went through traditional sales channels. The traditional farmer collected eggs once a day and sold them once a week to the general store or feed store in his home town. The store would send the accumulated eggs once a week by unrefrigerated slow freight to wholesalers in the nearest large town, who graded them and shipped them by unrefrigerated slow freight to whichever large urban area was paying the best prices, often more than a thousand miles away.

They say "nothing beats a fresh farm egg," but most farm eggs weren't fresh by the time they reached the consumer. So you can see why large farms that practiced end-to-end refrigeration gained the consumer's confidence, and how the elimination of a couple of levels of middlemen, plus economies of scale, helped price the factory-farm eggs so cheaply that people with small flocks couldn't compete.

4.2. Profitable Poultry

But, as with everything else, the egg and broiler markets are divided into commodity products and specialty products. The commodity products are produced in immense volumes and are sold at very low profit margins, while the specialty products are produced in small volumes and sold at high profit margins.

Becoming an independent producer of commodity eggs or broilers is impractical. The profit margins are more or less nonexistent.

The specialty markets are another matter. I have a free-range egg flock and my wife raises pastured broilers. Pasture-raised eggs and broilers have a lot of sales appeal and command high prices. At the time of this writing, the commodity price paid by large supermarkets for Grade AA Large eggs is about $0.75 per dozen. We are charging supermarkets $2.60 per dozen for free-range eggs and don't have enough eggs to go around. This is high enough to return good money for our labor in addition to paying the feed bill.

Hobbyists often sell their eggs for much less than this, which is fine, but practical farmers have to make a part-time or full-time livelihood from farming as an alternative to taking a job in town.

But we were talking about breed selection. Unless there is a very specific and profitable niche that demands some special kind of bird, the practical farmer has to choose highly productive breeds. This generally means that you have to use modern commercial hybrids.

"Heritage breeds" have received a lot of publicity recently, but they should be approached with caution. The reason they've been relegated to the "heritage" category is that they don't produce as well as modern commercial chickens. They not only produce less, they produce a *lot* less. The difference is more than enough to spell the difference between success and failure.

There are exceptions to this rule, of course, but they are usually based on a very specific market niche. It's best to test the limits of such niches carefully.

For example, Araucanas lay green eggs, and I had been asked about green eggs by many customers, who seemed eager to pay premium prices. As a test, I bought 100 green-egg pullet chicks to see if their had profit potential. This represented about 25% of the chicks I started that year. The results were disappointing. The pullets were late-maturing and laid poorly. I charged a fifty-cent premium for a dozen green eggs, but this wasn't enough to make the low-producing green-egg hens as profitable as my other hens.

Customers were interested in the green eggs, and a lot of them bought a dozen, but most of them only bought them once before switching back to our other eggs.

I made money selling green eggs, but I would have made more money if I had bought 100 brown-egg hybrid layers instead of the Araucanas. I didn't buy any more Araucanas.

4.3. Choosing Broiler Chickens

Commercial broilers are all hybrids, usually hybrids of four different strains. In other words, each of the four grandparents of a broiler chick is completely unrelated; they may be from different breeds.

Before the 1960s, most broilers were from dual-purpose breeds such as New Hampshires or White Wyandottes, or hybrids between two dual-purpose breeds, such as New Hampshire-Barred Rock hybrids. The Rock-Cornish hybrid eventually took over, combining the flavor of the Barred Rock mother with the meaty breast and plump, stubby drumsticks of the Cornish father.

The Rock-Cornish hybrid was slow to take over because both the Barred Rock and the Cornish had serious deficiencies as broilers. The Cornish was notorious for poor egg production in the females, poor fertility in the males, slow growth, and high mortality under broiler-house conditions. The Barred Rock, though a good layer, was slow-feathering, and was a second-rate broiler compared with the New Hampshire and other breeds. Year by year, starting in the Forties, the Cornish crosses improved, and they finally took over after 25 years of development.

Nowadays, the Cornish cross broiler carcasses look normal to us, while the carcass of a dual-purpose bird looks surprisingly comical, because it looks exactly like a rubber chicken!

Modern Cornish cross broilers grow quickly. Incredibly quickly. At slaughter age (6-9 weeks) they will weigh more than twice as much as a dual-purpose cockerel of the same age.

This fast growth can lead to severe health problems if there are any problems with your management. Even with excellent management, you will probably see growth-related health problems in a few broilers; at a guess, somewhere between 2% and 5%. Broilers approaching slaughter age can develop heart conditions, especially if they were chilled or had other problems in the brooder house. Leg problems that leave the broilers unable to walk are also fairly common.

Karen and I used to think that broiler chicks were harder to brood than dual-purpose chicks, but as our management improved, the difference disappeared, and we can't quite remember why we thought they were such a problem. With the improvement of our brooder-house technique, the problems in older broiler were greatly reduced.

You can buy slower-growing broiler chicks that still outperform traditional breeds. These are far less likely to have heart or leg problems since they don't grow fast enough to aspire to such illnesses. Most hatcheries sell "red broilers" and "black broilers," which are preferred in some ethnic markets where white-feathered birds are disliked. These happen to grow more slowly than ordinary white-feathered broilers. At least one hatchery (Privett) has a slow-growing white-feathered broilers.

There are a variety of fast-growing broiler strains on the market. I frankly can't tell one from another, though many growers have a favorite strain and will use no other.

4.4. Choosing Egg-Type Chickens

The profits of an egg flock are strongly dependent on how well the hens lay—much more so than people imagine at first glance. This is because the cost of keeping a hen is about the same, regardless of how many eggs she lays, but the returns are based solely on egg production.

For example, let us compare three flocks of 100 hens. The first consists of a breed that has never been of any commercial performance (for example, Brahmas, Silver-Laced Wyandottes, and Dominiques), which lays about 8 dozen eggs per year. The second consists of hens from a breed that was competitive fifty years ago, and lays 16 dozen eggs per year (for example, New Hampshires and White Wyandottes). The third is a modern brown-egg hybrid that lays 20 dozen eggs per year (for example, Production Reds, Black Sex-Links, Brown Sex-Links)[*]. In the example I've calculated what the sales, feed cost, returns over feed cost, and gross profit would be in a free-range operation like mine.

Flock	Doz. Eggs Per Hen	Sales @ $2.50/ doz.	Feed Cost per Hen*	Return Over Feed	Gross Profit**
Non-Commercial Breeds	8	$20	$8.75	$11.25	$5.05
1950s-Era Commercial Chickens	16	$40	$9.50	$30.50	$23.10
Modern Commercial Chickens	20	$50	$11.00	$39.00	$31.00
* Figured at $0.10/lb. ** Gross profit is figured as return over feed minus the cost of egg cartons ($0.15 per dozen) and the cost of raising a pullet chick to maturity ($5.00).					

Figure 1. Example of egg sales minus feed cost for three classes of hens

As you can see, the non-commercial breeds are non-starters, bringing in a gross profit per hen that is less than one-fourth that of the 1950s-era birds, and one-sixth that of modern commercial chickens.

The message is clear: non-commercial breeds and a profitable egg flock are mutually exclusive. Keeping them pushes your flock irretrievably into the "hobby" category, and you need to be aware that the "profits" probably won't even return the cost of housing and fixtures. There will be nothing left over to pay for your labor.

Why haven't all breeds been improved to commercial levels? It turns out that breeding for increased egg production is difficult and expensive. Such breeding activity has always been focused where it was most likely to pay off: on breeds that were high-producing to begin with, such as White Leghorns, Rhode Island Reds,

---

[*] In modern commercial confinement they would lay 26 dozen per year, but 20 dozen seems about right for free-range operations like mine.

and Barred Rocks. As long ago as 1900, these three breeds made up the vast majority of commercial laying hens. The other breeds had already been rejected as unsuitable for commercial production.

4.4.1. *Commercial Breeds*

Most commercial egg breeds are either hybrids between two breeds of chicken, or a strain cross between unrelated families within the same breed.

White Leghorns, Rhode Island Reds, and Barred Rocks still exist as commercial layers. However, each of these breeds has split into two categories: standard-bred chickens and utility chickens.

Standard-bred chickens are bred with the goal of producing offspring that match the description in *The Standard of Perfection*, the bible of poultry fanciers, which focuses almost exclusively on what "perfect" chickens would look like. Each breed has its own description. Unfortunately, being pretty enough to win a blue ribbon is unrelated to egg production.

Utility chickens are commercial chickens, which are bred for high production, regardless of what it does to their good looks. Utility chickens are unlikely to do well if entered in a poultry show.

Hybrid chickens are virtually always utility chickens, since there are no categories for them in poultry shows.

I have read some wild claims about utility breeds, claiming that they are no good for one reason or another. Some writers who don't like factory-farming will extend their criticism to include the chickens living on factory farms—guilt by association. This hardly seems fair. My experience is that utility breeds, taken as a whole, make superior farm chickens to standard breeds. There are a lot of utility breeds, each with its own traits. Some are more likeable than others, and they respond differently to a given set of management techniques.

Many of the most important characteristics of a good farm hen are variable within a breed. For example, Black Sex-Links from one supplier might be docile and friendly, while from another supplier they might bite your hands whenever you try to collect eggs, forcing you to wear gloves. Because of this, a detailed description of each breed is not feasible. However, it's probably worthwhile to list the contenders for commercial-breed status.

(Let me apologize in advance to everyone who feels that I've put their favorite breed in a less favorable category than they feel it deserves.)

4.4.2. High-Producing Commercial Strains

Brown Eggs

- *Production Reds* (a strain cross between two high-producing strains of Rhode Island Reds).
- *Black Sex-Links* (a cross between a Production Red and a Barred Rock).
- *Name-brand hybrids,* such as ISA Browns, Golden Comets, Cherry Eggers, and so on (usually a four-way cross where each of the chick's four grandparents comes from a different breed or strain of chicken).
- *Brown Sex-Links* (a fairly generic name for brown-egg sex-linked hybrid. They have the same general characteristics as the name-brand hybrids)

White Eggs

- *Commercial White Leghorns* (a strain cross between two or four high-producing strains of White Leghorn)
- *California Whites* (a cross between a California Gray and a White Leghorn)

Tinted Eggs

- *Austra Whites* (a cross between a Black Australorp and a White Leghorn), laying eggs midway in tint between brown and white. Some strains lay white eggs, not tinted eggs.

Green Eggs

- None.

4.4.3. Marginal Commercial Strains

Brown Eggs

- *Barred Rocks and White Rocks.* Some commercial strains of are very good. Because the Barred Rock was the most popular farm chicken in America, many more breeders worked with it than most other breeds. White Rocks are a white-feathered variant that is more or less interchangeable with the Barred Rock. These hens are calm, docile birds, though the roosters have a reputation for viciousness. Beware, though, that some strains are standard-bred, some are utility egg strains, and some are no doubt old utility broiler strains. Avoid Rocks unless you are assured that they are a high-producing egg strain.
- *Rhode Island Red.* Some individual Rhode Island Red strains produce almost as well as Production Reds, for reasons similar to the Barred Rock (the Rhode Island Red was the number two breed in America). Rhode Island Red strains are as variable as Barred Rock strains, so be careful only to consider strains that are advertised as having particularly high egg production.

White Eggs

- *California Gray.* This is a fascinating breed. Grays are auto-sexing, meaning you can tell the sexes apart at any age. They lay large white eggs, but are not panicky and flighty like Leghorns. They are larger than Leghorns but not as large as most of the brown-egg birds. (California Whites are midway between Leghorns and California Grays in temperament, and lay more eggs than California Grays, though not as large.)
- *White Leghorn.* Some non-strain-cross, standard-bred White Leghorns are pretty good, I'm told. They will behave just like the commercial birds (that is, they will fly early and often, and will be panicky.)
- *Brown Leghorn.* These lay pretty well by all accounts. In temperament and size they are the same as White Leghorns.

Green Eggs

- None

4.4.4. Non-Commercial Strains

Brown Eggs

- New Hampshires, Delawares, Orpingtons, Wyandottes, and Australorps are the also-rans in the breed list. At one time they were considered to be suitable as commercial layers, but this is no longer the case.
- All other breeds. Dominiques, Houdans, Games, Jersey Giants, Brahmas, Cochins, and all the other breeds not mentioned are not of commercial quality. Some were considered good layers 150-200 years ago, but that's about the best that could be said for them.

White Eggs

- All white-egg breeds not mentioned above should be considered questionable. High-laying strains of Minorcas and Anconas used to exist, but they haven't been used commercially for fifty years, and even then were not highly thought of.

Green Eggs

- Araucanas, Americaunas, Easter-Egg Chickens, and green-egg birds going by other names are not worth considering on a practical farm. Interesting birds, though.

Chapter 5. Brooding Basics

5.1. Why Brood Chicks?

A baby chick doesn't have the ability to maintain its own body temperature without an external source of heat. In natural brooding, this heat is provided by the chick's mother, a broody hen. The chicks nestle among her fluffed-out feathers and take advantage of both her body heat and the insulation afforded by the feathers.

Figure 1. Contented chicks sleeping around an old-fashioned oil-fired brooder.

Artificial brooding involves either an external source of heat to take the place of the hen, or a specially designed box so the chicks can huddle together and stay warm with only their body heat. This second method is what allows chicks to be shipped through the mail without a heat source. It is not very practical in other contexts.

As the chicks become older, they require less and less heat. When they no longer require heat at all, the brooding period is at an end. This can range from two to three weeks in hot weather to five or even six weeks in cold weather. Broiler chicks are fast-growing and require heat for a shorter time than other breeds.

Chicks grow at a tremendous pace. As they grow, they require more food, more water, more ventilation, and more space. They grow so fast that they can go from needing supplemental heat to stay warm to needing supplemental ventilation to stay cool in just a few days. They will outgrow at least one and possibly two sets of feeders and waterers during this period. They also go through marked behavior changes during the brooding period.

5.2. Heat

There is less variety in brooders today than there was fifty years ago, and most offerings are geared to either very large or very small flocks. A particularly regrettable loss is that of the 200-250 chick brooders which were once a mainstay of American farmers. I will describe how you can make your own insulated electric brooder in a couple of hours for about $20.

The heat source most appropriate for a given flock depends mostly on the size of the brooding operation. For most of us, electricity is the right fuel until we start brooding thousands of chicks at a time, at which point propane (or natural gas) becomes the right fuel. Both electric and propane brooders are simple and reliable, but propane is generally cheaper. Big commercial operations that brood tens of thousands of chicks at once almost always use propane.

Most brooders are either infrared brooders or warm-air brooders.[*] Infrared brooders typically use a heat source to warm the chicks by shining hot rays on them, rather than by heating the air around the chicks. Warm-air brooders warm the air, which warms the chicks.

Young chicks lack the ability to regulate their own body temperatures internally, but they can do it externally by shifting closer

---

[*] There are also exotic types, such as radiant floor brooders, that will be discussed briefly in a later chapter.

to the heat source when they feel cold, or further away if they feel hot. It's important that they be allowed to do this; a room with absolutely uniform temperature doesn't work very well (and is hard to achieve, anyway). There needs to be a place where the chicks can warm up and a place where they can cool down. They can go to one or the other, or stay somewhere in between, according to how hot or cold they feel at the moment. This happens even in thermostatically controlled brooders. Thermostatic control of the brooder isn't really necessary, since the chicks will be moving around to adjust their personal comfort levels even if the brooder conditions are constant. Warm-air brooders generally have thermostats, but infrared brooders generally do not.

Most brooders include a canopy or "hover" that helps hold the warm air close to the floor, rather than letting it rise up to the ceiling. This lowers energy costs.

5.3. Adjusting the Temperature

The following quote from almost a century ago describes things as well as I could:

> The temperatures given for running brooders vary with the machine and the position of the thermometer. The one reliable guide for temperature is the action of the chicks. If they are cold they will crowd toward the source of the heat; if too warm they will wander uneasily about; but if the temperature is just right, each chick will sleep stretched out on the floor. The cold chicken does not sleep at all, but puts in its time fighting its way toward the source of heat. In an improperly constructed or improperly run brooder the chicks go through varying processes of chilling, sweating, and struggling when they should be sleeping, and the result is puny chicks that dwindle and die.[*]

In adjusting the temperature, a thermometer gives a good starting point, but only a starting point. A thermometer can't factor in the effect of floor drafts. Also, different parts of the brooder are at different temperatures.

---

[*] Milo Hastings, *The Dollar Hen*, 1909, The Arcadia Press, p. 65.

The usual advice is to set the temperature at 90-95 °F at a point two inches off the floor and at the edge of the hover. This advice is hard to follow in brooders that don't have a hover.

Infrared light interacts strangely with thermometers. Placing a piece of black electrical tape on the bulb of the thermometer will give more consistent readings.

What most writers forget to mention is that every brooder should also have a region that's too hot for long-term comfort, so a cold chick can warm itself up in a hurry. It is easier to test this with your hand than with a thermometer. If the hottest part of the brooder isn't uncomfortably warm after you've kept your hand there for a minute or so, you probably don't have enough heat.

Once the chicks are installed, their own behavior makes a thermometer unnecessary. If you use a thermostatically controlled brooder, it only comes into play once a week, when you reduce the temperature by five degrees. After doing so, you should check after a few hours to make sure the chicks aren't acting cold.

5.4. Chicks and Light

Chicks are attracted to light. When baby chicks are placed in a brooder house, they don't know which parts are warm and which are cold. But if the heat source is also the brightest part of the brooder house, the chicks will be attracted to it.

Some brooders, such as infrared lamp brooders, give off light in their normal operation. Ones that don't are generally provided with light bulb sockets for "attraction lights." After about three days, the chicks will have learned where the heat is, and will no longer need extra light as a beacon.

If there are patches of sunlight in the brooder house, the chicks will be attracted to them, and may not move when the sun goes down. If they don't go back to the brooder, they may chill or smother as they huddle together for warmth. Small patches of sunlight are more troublesome than large ones, since more chicks will try to cram themselves into a small patch than can be accommodated there. Lights under the brooder will help convince the chicks to head home at sundown.

Chicks can't eat or drink in the dark. It's a standard practice to leave the lights on continuously for the first three days, so the chicks have more opportunity to find the feeders and waterers. This light should also be relatively bright. Once they know where everything is, they don't need much light to get around, but when everything is new and strange, they are better at figuring out the world around them if it is well illuminated.

5.5. Drafts, Ventilation, and Wind Chill

Practically every book on poultry insists that the brooder house has to be "well-ventilated but free from drafts," but they rarely give any hint of what they are talking about. Sometimes I suspect the writers didn't know, either.

The issue revolves around floor drafts and wind chill. Any breeze at chick level (that is, within a few inches of the floor) will cool the chick and reduce the effective temperature. If the temperature at chick level is too hot, a breeze can provide much-needed cooling. If the temperature is just right or on the cool side, a cooling breeze can be disastrous. Drafts high above the chicks are a problem only if they stir up a draft a chick level.

5.5.1. Curtains

Sometime curtains about four inches high are attached to the hover to hold in the hot air and stop drafts. It used to be common for gas and electric brooders to have curtains all the way around the perimeter. I don't know of any currently manufactured units that have curtains, but you see them on old units at farm sales. Slits in the curtains made it easier for the chicks to get past them.

Curtains make it difficult for a chick to find a comfortable place to sleep if the brooder is a little too warm. In a brooder without curtains, the chicks will find a comfy spot a short distance outside the hover, but the curtain eliminates the gradual transition from inside to outside. Also, curtains make it almost impossible to see under the brooder. You have to lift it to check on the chicks.

If the brooder house is cold, it may be necessary to use curtains in spite of their disadvantages.

Often the drafts in a brooder house come from a single direction—usually from the front of the house. A curtain on just one side of the brooder will often eliminate the draft problem without buttoning up the brooder too much.

5.5.2. Draft Guards

The usual method of providing comfortable heat levels for the chicks is to surround the brooder with a draft guard for the first few days, to eliminate the slightest hint of a floor draft. This reduces the amount of heat that has to be provided and prevents one from having to worry about wind chill. This guard is typically a ring of cardboard, 12-18 in. high, that surrounds the heat source. There needs to be enough space for the chicks to be able to escape the heat of the brooder if they feel hot, and plenty of room for feeders and waterers. I prefer aluminized bubble-wrap insulation to cardboard, since it lasts longer, is washable, and holds heat better.

Figure 2. A cardboard draft guard surrounding an old-fashioned electric brooder.

As the chicks become larger, they will scratch around in the litter and will eventually open tunnels under the draft guard. The draft guard should be removed before this happens, because escaped chicks may not be able to find their way back to the brooder and can die in the cold. If the chicks start escaping, shore up your defenses or remove the draft guard immediately.

In cold weather, a good draft guard is absolutely essential for the first several days, maybe the entire first week. The draft guard is expanded day by day to give the chicks more room.

In hot weather, where the chicks are more likely to become overheated than chilled, the draft guard is either eliminated or made out of ¼" or ½" hardware cloth. This allows the floor draft to cool the chicks without allowing them to wander too far away from the brooder.

5.5.3. Ventilation

There's an old poultrykeeping proverb: "The best chicks come out of the sorriest houses." This is because it's human nature to keep the brooder houses sealed up too tightly, leading to a build-up of moisture, ammonia, and pathogens. A tumble-down chicken house with doors and windows that don't seal properly and gaps in the walls won't have this problem.

When the chicks are very small, they are very susceptible to chilling, and it's best to button up the house pretty tightly to minimize airflow. Even with a draft guard around the brooder, strong breezes in the house will set up turbulence that will be felt at chick level.

But as the chicks get older, the moisture from their respiration and manure increases dramatically. Pathogens like roundworms and coccidiosis that can't survive in a dry house flourish in a damp one. Moisture can't escape from a tightly sealed house. You have to increase ventilation as the chicks get bigger.

Ammonia from the chicks' manure is also a health hazard to the chicks. Additional ventilation will prevent it from accumulating to dangerous levels.

Ventilation needs to be increased long before the brooder is turned off. You will know it's time to increase ventilation if the litter becomes damp or caked or if there's an ammonia smell in the house. But it's always tempting to put it off through fear of chilling the chicks. If your house is sufficiently badly sealed, your chicks will escape most of the consequences of this procrastination.

In cool weather, ventilation near the ceiling is all that's required. In very hot weather, a floor draft may be necessary to cool the chicks. Double-hung windows are handy for providing both kinds of ventilation.

In cool climates, all the ventilation is provided at the front of the house. In hot climates, cross-ventilation is necessary. At a minimum, vents should be installed at the back of the house, just under the roof. These should have shutters so they can be closed in cool weather.

Modern large-capacity poultry houses have forced-air ventilation, but they are subject to catastrophic losses if the power fails on a hot day. To avoid this, they add automatic backup generators in addition to the fans. This is not an expense I would willingly undertake unless I had tens of thousands of chicks.

5.6. Warm-House vs. Cold-House Brooding

There are two basic methods of artificial brooding: the warm-house method and the cold-house method.

5.6.1. Warm-House Brooding

In the warm-house method, the brooder house temperature is kept above some minimum level, such as 70 or 85 °F. Depending on the level of insulation in the house, this may be achieved through the heat given off by the brooders themselves, or there may be a central heating system.

The commercial poultry industry uses warm-house brooding because it's more foolproof than cold-house brooding, since the chicks won't get badly chilled even if they wander off and become lost. The method is suited to brooding enormous flocks at once, where it's hard to keep an eye on every chick. Brooders used in warm-house brooding are simple and uninsulated. Typically, the waste heat from the brooders goes a long way toward keeping the house warm.

Warm-house brooding calls for a purpose-built brooder house with substantial amounts of insulation, or maybe a powerful brooder burning very cheap fuel.

Some people achieve warm-house brooding by brooding chicks inside their homes. I have done this myself. I cannot recommend the practice, partly because of the flies, dust, and smell, and partly because of the small but real danger of fire wherever chicks are brooded. Chicks should be brooded in outbuildings, preferably ones that aren't particularly valuable, and which have metal roofs.

5.6.2. Cold-House Brooding

In cold-house brooding, only the area under the brooder itself is kept warm. The rest of the house is left to take care of itself. This method is well suited to the semi-improvised, uninsulated brooder houses that many of us use, and when brooding small flocks. It is difficult to justify the expense of keeping a large shed at 85 °F if you are only brooding 25 chicks.

Cold-house brooding is particularly cost-effective with insulated brooders, although insulated brooders are no longer manufactured and have to be home-made. Hanging infrared heat sources are more commonly used, since they work by heating the chicks directly with infrared rays, not by heating the air under the hover. Infrared brooders generally don't even have a hover. There are a variety of infrared heaters on the market that are suitable for brooding poultry.

Brooding at sub-freezing indoor temperatures is perfectly practical with cold-house brooding, provided that floor drafts are controlled and the feed and water are kept within the warm area near the brooder.

5.7. How Much Heat?

The brooder should be fired up at least 24 hours before the chicks arrive. This will give you time to run into town and buy parts if you have to. If you use a thermostatically controlled brooder, have a spare thermostat wafer (or, better yet, a spare thermostat assembly including the microswitch) on hand. If you use heat lamps, have a spare bulb ready.

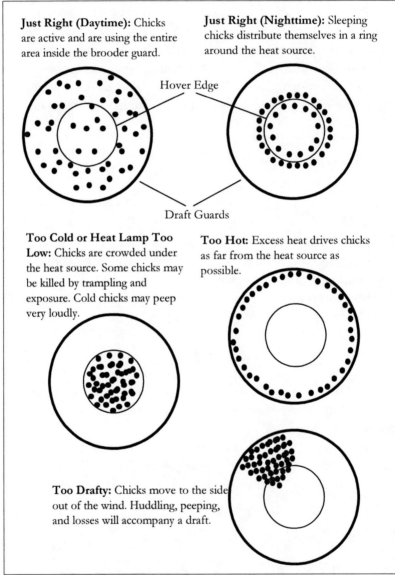

Just Right (Daytime): Chicks are active and are using the entire area inside the brooder guard.

Just Right (Nighttime): Sleeping chicks distribute themselves in a ring around the heat source.

Hover Edge

Draft Guards

Too Cold or Heat Lamp Too Low: Chicks are crowded under the heat source. Some chicks may be killed by trampling and exposure. Cold chicks may peep very loudly.

Too Hot: Excess heat drives chicks as far from the heat source as possible.

Too Drafty: Chicks move to the side out of the wind. Huddling, peeping, and losses will accompany a draft.

Figure 3. Using the chicks' activity to monitor brooder temperature.

The floor in the heated area should be warm and dry to the touch before the chicks arrive.

If you have a thermostatically controlled brooder, use a thermometer to verify that the temperature is set correctly. This generally means having a temperature of 90-95 °F, one or two inches off the floor at the edge of the brooder. However, if you have a user's manual for the brooder, take the measurement the way they recommend, not the way I recommend.

You can probably gauge an infrared heat source better with your hand than with instruments.

Most infrared brooders don't have an intensity control. To make a larger, cooler heated area, the heat source is raised higher off the floor. To make a smaller, hotter area, the heat source is lowered.

Once the chicks have arrived, their behavior is the best guide to whether the heat is right or not. See Figure 3.

5.8. First Water

Newly hatched chicks should be given water as soon as they are placed in the brooder.

Newly hatched chicks are very tiny, and can become soaked in waterers designed for adult hens. If this happens, they will generally die from the resulting chilling. This risk goes away after a few days, but it's a serious danger at the beginning. I have found that the chick founts in one-gallon through five-gallon sizes are dangerous for the first three days or so, in spite of being labeled as suitable for day-old chicks.

I recommend quart-jar waterers for baby chicks. The waterer base (your choice of galvanized steel or red plastic) screws onto a standard narrow-mouth quart canning jar. You can also buy translucent plastic jars at the feed store along with the waterer base, but I don't recommend it. The chicks seem to know the glint of water by instinct, and shiny glass jars will attract a thirsty chick where a translucent plastic ones will not. Once they're attracted to the waterer, they'll peck at the glass a couple of times, and then find

Figure 4. Chicks around a quart-jar waterer.

the water. Besides, it's easier to see how much water is left in a glass jar than in a plastic one, and glass jars are cheaper.

Some authorities suggest that you put marbles in the water for extra glint, but I don't think this is necessary if you use glass jars. Others suggest that you dip the beak of each chick into the water to stimulate drinking and teach some of the chicks where the water is. The chicks key on each other in addition to the glint of water, and if they see other chicks drinking, they'll come over and start drinking, too. I used to do this, but I gave up after noticing that the chicks always start drinking on their own right away without my help.

Many books say to put the waterers flat on the floor. Maybe this will work for you, but it never has for me. The waterers tend to lean like the Tower of Pisa and leak out all their water, and the chicks kick litter into them. I either put them on a scrap of lumber (a scrap of 1x4 or 2x4 works okay, but a 1x6 is better) or on some

kind of grid or mesh support that sits flat on the ground. Some
people build special platforms for this, using ½" hardware cloth
on a wooden frame.

5.8.1. Use warm water

The first water you give the chicks should be warm. Baby chicks
can arrive dehydrated and partly chilled, and a big drink of ice-
cold water is the last thing they need. When it comes time to refill
the waterers, cold water can be used.

5.8.2. Additives

There are many additives that people put in the water for baby
chicks. Probably the best-proven additive is ordinary sugar at the
rate of two cups of sugar per gallon of warm water. This is given
in the chicks' first water only. When it comes time to fill up the
waterers again, empty the waterers and refill them with tap water.
The idea is that chicks arrive on the farm hungry and dehydrated,
and you can attack both problems at once with sugar-water, and it
has been demonstrated to reduce chick mortality.[*] I almost always
use sugar water with my chicks.

Some hatcheries recommend that you put antibiotics in the
first water (2 teaspoons of Aureomycin or Terramycin per gallon
of water). Others recommend that vitamin/electrolyte mixtures
be used routinely for the first three days. I don't use anything but
sugar in the chicks' water.

It's also quite common for farmers to put low levels of sanitizers in the water to reduce the spread of disease (which is often
spread via waterers). I don't think this is appropriate for the
chicks' first water. The waterers should be cleaned before use, and
sanitized if you want, but they should be filled with water that
doesn't have lots of sanitizers it. Chickens will refuse to drink
water with high levels of chlorine in it, for example, until they've
had several days to get used to it. We can't afford this kind of time

---

[*] North & Bell, *Commercial Chicken Production Manual*, 4th edi-
     tion, p. 240.

with newly hatched chicks. (City water should have low enough levels of chlorine that there isn't a problem.)

5.9. First Feed

North and Bell recommend witholding feed for three hours after placing the chicks in the brooders.[*] Other authorities say you ought to provide feed right away.

Figure 5. Baby chicks feeding out of a home-made wooden feeder

5.9.1. First Feeders

Newly hatched chicks have a strong instinct to scratch around on the ground for their feed. It can take them several days to learn to eat properly from ordinary feeders. Thus, to help them in the all-important first days of life, it's best if we provide them with feed on the ground. It is generally fed in shallow trays or egg flats. I like plastic cafeteria trays for this task. Egg flats are also good. Provide

---

[*] North & Bell, *Commercial Chicken Production Manual*, 4th edition, p. 240.

one plastic cafeteria tray or one egg flat of feed for every 50 chicks.

A traditional first feeder is the box the chicks arrived in, with its sides cut down to about one inch high. Another solution is to use the lid of the box, though the ones I've been receiving in the last few years aren't very suitable for this. Home-made wooden trays with plywood on the bottom and a rim of lath or 1x2 work fine.

Some people put sheets of newspaper or kraft paper on the floor and pour the feed onto that. Such paper should be rough, to give traction. If you use newspaper, use only a single thickness. Thick wads of newspaper will become slick with manure. Single thicknesses will be ripped to shreds before this happens.

Figure 6. Feeder made by cutting down a chick box

Floor-level feeding is continued for 5-7 days. This method of feeding is wasteful in terms of the percentage of feed that is lost, but the total loss is small when measured in pounds. The waste is minimized if relatively small amounts are fed at a time, with fresh feed added several times a day. Chicks can tell the difference between fresh and stale feed.

Ordinary chick feeders can be introduced at the same time as the flats, or added within a few days of the chicks' arrival, so they have a chance to get used to them before the first feeders are discontinued. Chicks are not fast learners, so new equipment should always be introduced before they are forced to rely on it.

5.9.2. Chick Starter and Chick Scratch

The first feed should generally be chick starter feed: a balanced chick ration that comes in the form of small "crumbles" (crushed pellets). Sometimes chicks have trouble with "paste-up," where manure dries to their rear ends, giving a case of constipation that will sometimes kill them. Giving them a diet of chick scratch (mixed, finely cracked grains) for two days before giving them chick starter will generally prevent this. Paste-up is generally a symptom that it's too cold or too drafty under the brooder, and fixing this problem will do a lot more good than anything you can accomplish by changing the feed. But feeding scratch grains for the first two days is not harmful to the chicks. A compromise is to feed chick scratch on the floor and chick starter in the feeders. The chicks like chick starter a lot better than they like scratch grains, so this ought to encourage them to learn to use the feeders.

5.10. Second Feeders and Waterers

You can have the second feeders installed and filled at the same time as the first feeders, or install them after a day or two. The second waterers should be withheld for a couple of days, since it takes that long for the chicks to become resistant to getting soaked in the waterers.

Chick feeders take the form of troughs or tube feeders. I generally use tube feeders hung from the ceiling, since their height can be adjusted as the chicks grow. If you use troughs, you'll need three different sizes by the time the chicks are grown (chick troughs, pullet troughs, and hen troughs), two of which will be used in the brooder house. Only two sizes of tube feeder need be used. The small 15-pound capacity tube feeders will do for the brooder house, and the larger ones with much deeper feed pans prevent larger birds from wasting feed.

When using tube feeders, start out with the pans flat on the floor. Raise them after a week or so until the height of the feed is roughly level with the chicks' backs. If they have to strain upward to get feed, the feeder is too high.

Most kinds of poultry waterers can be used for second waterers, although hen-sized trough and bucket waterers are too big for use in the brooder house. Waterers are covered in Chapter 12.

5.11. When to Start Chicks

Retail hatcheries tend to hatch commercial breeds (hybrid layers and Cornish-cross broilers) year-round, while their standard-bred chicks are available only from late winter to late summer. This is because farmers with small commercial flocks buy chicks all year, but hobbyists and backyarders don't. Thus, if you are interested in standard-bred chicks, you have about a 6-8 month window in which to buy them (roughly February through August).

Chicks are easiest to brood when the weather is not too hot and not too cold. April and May are the traditional times to brood chicks, which in most parts of the U.S. represent times where the temperatures are reliably above freezing but the days are not yet hot enough to be dangerous.

A traditional goal of egg farmers is to have the new flock laying well by the first of November, because this will compensate for the declining production of last year's hens, whose production peaks in April and May and declines steadily thereafter, reaching its minimum sometime between November 1 and the end of the year. Depending on how fast the chicks mature, the new flock will have to be hatched somewhere between and April 15 for traditional breeds and June 15 for commercial egg breeds.

However, if you sell a lot of eggs at farmers' markets or other seasonal outlets, sales tend to peak during July, August, and September, when everyone has fresh produce on their minds. To get the pullet flock laying by July 1 calls for commercial egg-type chicks by mid-February.

There is also something to be said for starting chicks in the fall. September is a good month for starting chicks if the climate isn't too hot, and October is good if the climate isn't too cold. Fall chicks will over-winter perfectly well if they are well-grown by the time really cold weather sets in—say, eight weeks old. They will start laying in the spring at about the same time as the production of the old hens starts to pick up. Many small producers have few outlets for egg sales in the winter, and disposing of all their hens in the fall and brooding a new batch of chicks may make more sense than overwintering the hens. The growing chicks will eat only about half as much as the same number of full-grown hens.

By brooding several batches of chicks per year, you don't need as much brooding equipment. I am brooding all my chicks in two tiny brooder houses that hold only 75 chicks each. By brooding four times a year (February, April, June, and October), I raise 600 chicks a year.

Chapter 6. Overhead Heat-Lamp Brooders

You can make an effective brooder by taking a 250-watt heat lamp and suspending it 18-24 in. off the floor of the brooder house, with the bulb pointing straight down. Baby chicks are naturally attracted to light, so they will instantly be drawn to the circle of light and heat provided by the lamp. If they feel too warm, they will drift towards the edge of the beam, and if they feel too cold, they will move to the middle.

Figure 1. A typical brooder lamp. The wire guard on the front keeps chicks from flying into the bulb, and also causes the unit to tip onto it side if it falls to the floor, making a fire unlikely.

The number of chicks that a 250W heat lamp can support depends on the room temperature. Based on experiment station studies[*], at a 50 °F average room temperature, it can warm 75 chicks. For every degree above 50 °F, add one chick. For every degree below, subtract one chick. For example at 80 °F average room temperature, a bulb can warm 105 chicks, while at 20F average room temperature, a bulb can warm 45 chicks.

Room Temp,°F	Number of Chicks Per Bulb	
	125W Bulb	250W Bulb
-20	2	5
0	12	25
20	22	45
40	32	65
60	42	85
80	52	105

Figure 2. The maximum number of chicks per heat lamp varies with room temperature

The most extreme example of cold-weather brooding that I'm aware of was a demonstration in 1950, where a flock of day-old chicks was brooded inside a walk-in freezer at -10 °F. Brooder heat for the 25 chicks was provided by four overhead heat lamps. It was so cold in the freezer that ice formed on the waterers on the side facing away from the bulbs, but the chicks were perfectly comfortable.

The height of the bulb is important. With a 250W bulb, start with the bulb 18-24 in. off the floor. With a 125W bulb, start with the bulb about 13-17 in. off the floor. It's hard to be precise since the width of the beam varies by manufacturer and bulb type.

---

[*] From Purdue University, as reported by *Hatchery & Feed* in April, 1952.

After the chicks arrive, check to make sure that all the chicks fit within the beam with room to spare. If you see the chicks huddling and struggling in a solid mass inside the beam, the lamp is too low—they don't all fit inside the warm area, and have to struggle to keep from being ejected into the cold. If the lamp is too high, they will be too cold and they may peep very loudly. They will huddle, but not with the constant struggle you get when the bulb is too low.

6.1. Other Infrared Brooder Types

Not all overhead infrared brooders use the simple light fixture shown in Figure 1. There are at least two manufacturers of four-bulb fixtures with a thermostat that controls two of the bulbs, leaving the other two on all the time. Presumably, a four-lamp brooder will handle four times as many chicks as a one-lamp brooder.

Figure 3. A four-bulb brooder. Two lamps are thermostatically controlled; two are always on. Such a unit is good for roughly 300 chicks with 250W bulbs or 150 chicks with 125W bulbs.

Other infrared brooders use quartz tubes instead of heat lamps. They are more expensive than fixtures for heat lamps, but the quartz tubes last almost forever.

6.2. Use Two Bulbs or New Bulbs

With any lamp brooder, there's a chance that the lamp will burn out. Many people take this risk, but I always use a minimum of two lamps for each group of chicks. If one lamp burns out, all the chicks will end up under the remaining one, and it's likely that no chicks will be lost.

This doesn't require extra electricity because you can use smaller bulbs; for instance, two 125-watt heat bulbs instead of one 250-watt bulb. When using 125-watt bulbs, the bulbs should be hung a little lower, with the minimum height being about 13 in. off the floor instead of 18 in.

Figure 4. A homemade twin-lamp brooder. Note the use of 1x2 in. welded wire for lamp guards.

Some people like to use brand-new bulbs when starting a new batch of chicks, on the grounds that an old bulb may be ready to burn out. A six-week brooding period is about 1,000 long, which is the lifespan of an ordinary bulb. The average life of most reflector floodlights is 2,000 hours these days, and most heat lamps are rated for 5,000 hours or more. This implies that starting each year with a new set of bulbs might be all the safety precaution we need.

6.3. Safety

It's best to take a few pains when hanging the lamps. They should be secure but easily adjustable. Lightweight chain is good for this, with the lamp housing hanging from an S-hook.

Use fixtures designed for use as brooders. These have porcelain sockets and some kind of guard over the front to reduce the chance that the litter will be set on fire if the lamp falls to the floor. Every feed store carries them, at least in the spring. Hardware stores generally carry only an inferior grade of clamp light. Not only are these worse in every way than a real brooder fixture, but they tend to cost more!

Except for broiler chicks, which don't fly much, the chicks will start flying around and banging into things when they are a few weeks old. You will be amazed at what they can knock down. They can certainly knock down an imperfectly installed heat lamp.

The bulbs are very hot all over, including at the base, and this makes the sockets very hot. This is why it's important to use porcelain sockets.

Never try to unscrew (or tighten) a heat lamp when it's still hot. The glue that holds the lamp base to the glass is very weak when hot, and you'll twist the lamp right out of its base, often shorting it out in the process. Wait for the lamp to cool down first. And it's best to use gloves or a cloth when removing the bulbs, even if you think they've cooled enough.

6.4. Bulb Type and Wattage

Bulb choice can be important. The traditional choice is the 250W heat lamp, but heat lamps are also available in 100W, 125W, and

175W sizes. Heat lamps are either clear or red. The clear bulbs cast a lot more light and the chicks will be more active if you use them. The red bulbs discourage activity, but cost quite a bit more.

I think that it's a good practice to use clear bulbs for the first three days at least, because the brighter light is better at attracting lost chicks to the heat, and also makes it easier for them to see the feed and water. After they learn the ropes, they can function in the dim light of a red bulb.

If your chicks show any signs of feather-picking, toe-picking, or cannibalism during brooding, switching to red bulbs and reducing the amount of light coming into the brooder house will help. Otherwise, the choice of color probably doesn't make a lot of difference. I have never had problems with cannibalism during the brooding period, so I don't hesitate to use clear bulbs. At the first sign of trouble, though, I'd switch to red bulbs.

Ordinary reflector floodlight bulbs can be used instead of heat lamps. You can use either indoor or outdoor bulbs, which will give you a wide range of wattage, from about 30 watts to 150 watts. I have used 150W floodlights instead of heat lamps, and they worked fine. Heat lamps have a longer life than any of the alternatives, and this reduces the chance of burnout.

Some bulbs are much sturdier than others. Outdoor floodlight bulbs won't crack even if water is sprayed on them when hot. You can also get heavy-duty heat lamps which are pretty much indestructible. Ordinary heat lamps and indoor floodlights are lightly constructed. In spite of this, I use ordinary heat lamps and indoor floodlights without hesitation, partly because I always use two bulbs.

6.5. Draft Guards for Infrared Brooders

Raising chicks successfully in a cold room with overhead lamps requires good draft control, at least for the first week or so. Any drafts at floor level will chill the chicks and render the lamps ineffective. An effective draft guard is absolutely essential to the success of overhead heat-lamp brooders.

Because chicks have a tendency to pile into corners and suffocate each other when they are frightened, a circular draft guard is best (since it doesn't have any corners). Barring that, a guard of six or eight panels is better than a rectangular one.

Figure 5. A brooder house in the Fifties using dozens of four-bulb brooders. Note the cardboard draft guards and the feeders made from cut-down chick boxes.

A series of cardboard panels stapled, glued, or taped together will make an excellent draft guard, which can be used for multiple batches of chicks until it gets dirty or damaged. (I recommend that you not hold it together with duct tape, which seems to have a special affinity for the down of baby chicks. A chick that becomes stuck to the draft guard won't be able to get free, and might die of exposure.) Some people make hinged plywood draft guards.

My favorite draft guard material is aluminized bubble insulation, sold under various trade names (I use TekFoil). This stuff is basically bubble-wrap with a layer of aluminum foil bonded to it on both sides. The aluminum layer reflects heat very well, and the bubbles act as an insulator. It is easily cut with scissors and can be taped or stapled together as if it were cardboard. Because it reflects both heat and light, the area within such a draft guard is both warmer and brighter than with a cardboard guard, and this will help. Aluminized bubble insulation is also waterproof and can be hosed off and used again.

The draft guard has to be especially tight at the bottom. The chicks will pop through any little gap. Once out, they often can't find their way back to the brooder. Heaping litter against the outside of the draft guard works fairly well, but you have to keep an eye on it and build it up again if the chicks start scratching their way to freedom.

The draft guard should be large enough that all the chicks can get entirely outside the beam on a warm day. You also need room for feeders and waterers on the fringes of the beam. A draft guard four feet in diameter is the smallest I'd use for a single-lamp setup.

As the chicks get larger, the draft guard becomes less necessary and gets more in the way. Some people expand the draft guard day by day. I normally don't do this, but then my brooder house is only eight feet square, so not much expansion is possible in any event. I remove the guard when it becomes burdensome, usually when the chicks start to dig underneath it and become lost on the other side. Another alternative is to cut a few doorways in the draft guard to allow the chicks access to the other side. While they get lost if they have to duck under the guard, they can find a doorway more easily.

To gauge how well the chicks are doing in the brooder, check on them at the coldest part of the day. If they are huddled in the middle of the light beam, they're cold, and you should add another lamp. If they're spread out in a ring around the edge of the beam, they're warm enough. If they're actually outside the beam, you can probably dispense with the draft guard. One rule of thumb is to raise the lamp two inches every week.

Some people use stock tanks or plastic wading pools as combination brooder-house floor and draft guard, putting a few inches of litter in the bottom and suspending a heat lamp over the center. This works very well, but don't use automatic waterers inside such a structure, because the chicks may drown if the water valve gets stuck open—something that happens all too frequently with automatic waterers. The problem with things like wading pools is that the chicks are soon big enough to fly over the top, and you have to be clear in your mind how you are going to proceed once this happens. Removing a wading pool or a stock tank full of chicks from the brooder house is not easy.

Chapter 7. Insulated Heat Lamp Brooders

A different kind of infrared brooder was developed in 1940 by the Ohio Experiment Station. This used heat lamps inside an insulated hover. Instead of pointing down, the heat lamps were mounted horizontally. I have used the Ohio brooder design for years and have been very happy with it. It uses about one-third the electricity per chick as overhead infrared lamp brooders and is easy to make.

In this chapter I am going to reprint the original Ohio Experiment Station write-up of the brooder. In the next chapter I will discuss it in more detail

Figure 1. Week-old chicks under one of my insulated heat-lamp brooders.

The following material was originally published in January, 1942 as *Special Circular 63* from the Ohio Experiment Station, under the title of "New Electric Lamp Brooder" by D. C. Kennard and V. D. Chamberlin. The text and illustrations are from the original. The footnotes are mine.

New Electric Lamp Brooder
D. C. KENNARD AND V. D. CHAMBERLIN

The new electric brooder to be described was designed and first used by this Station in October 1940. During the meantime, five of these brooders have been in almost continuous use. They have been used successfully for starting and brooding chicks throughout the year and for summer brooding of poults. This type of brooder was designed and is operated upon the basic principle that chicks or poults can be depended upon to adapt themselves readily to their heat and air requirements when ample heat and air are provided. This contention has been substantiated by the extensive use of these brooders throughout the year under widely varying conditions. In all instances, satisfactory results were secured with these simple, inexpensive brooders. At no time was there noticeable evidence of a need for thermostatic heat regulation, additional ventilation, or other items that would make these brooders more complicated and more expensive.

The new electric lamp brooder –

- Involves a minimum use of metals needed for war purposes.[*]
- Weighs about 30 pounds without insulation material.
- Accommodates 150 to 250 chicks when made 4 by 4 feet or 250 to 300 chicks when made 4 by 6 feet.
- Is operated on the basis of the behavior of and comfort of the chicks rather than thermostatic heat control or temperature shown by thermometer. Thermostatic heat control is unnecessary, since the chicks readily adapt themselves to their heat requirements and comfort in a brooder of this kind. A thermometer is misleading rather than helpful, since the ordinary

---

[*] This was published one month after Pearl Harbor.

thermometer can not be depended upon to indicate the radiant or infrared heat requirements of chicks or poults.
- Has a wide range of heat supply for special brooding requirements throughout the year.
- Requires no curtains during usual brooding conditions. In severely cold weather, curtains may be needed to conserve heat or prevent floor drafts; otherwise, curtains should not be used.

Figure 2. A 4 by 4-foot hover. Note 4-inch space on top for insulation material

Electric lamps have recently become available which offer new opportunities for brooding chicks and baby turkeys. These lamps are available in two types, 150-watt projector or reflector spot or flood lamps and 250-watt R-40 Bulb Drying Lamps, all of which project infrared or radiant heat rays, as well as light rays. The projector lamps [outdoor floodlights] are made of heavy glass and can be subjected to cold, rain, or snow when burning, whereas the less expensive reflector lamps, [indoor floodlights] made of thin glass, are liable to crack if subjected to water while burning. The projector and reflector lamps have a life rating of 1,000 or more hours, and a longer life can be secured by using 120-volt lamps on a 110- to 115-volt circuit. The 120-volt lamps generally serve for two brooding periods. The 250-watt R-40 Bulb Drying Lamps [infrared heat lamps] have a much longer life rating, 5,000 or more hours.

Figure 3. Some of the chicks take to the top of the hover.

The satisfactory use of such lamps for converting batteries without heating elements into battery brooders suggested using them for floor brooders. In both types of brooders, the lamps were placed in a horizontal position to project the heat and light across the hover rather than downward.

The floor brooding hovers designed and used extensively by the Ohio Agricultural Experiment Station are simple, inexpensive, and easily made of plywood or pressed wood. The sides are 12 inches wide and extend four inches above the top to provide ample space of the fine litter-insulation material (fig. 2). Desirable insulation materials are finely ground corncobs, shavings, sawdust, or fine peat moss. With this type of hover, unlike most, the chicks are encouraged to roost on top of the brooder. After the first 2 weeks, they take to the top during the daytime and thus leave more room for those remaining on the floor (fig. 3). The bottom edge of the hover is 4 inches above the floor. Side curtains can be used when needed during severely cold weather. If there are floor drafts, a curtain can be used on the one or two exposed sides.

The hover may be made 4 by 4 feet for 200 to 250 chicks or 4 by 6 feet for 250 to 300 chicks. The lamps are placed in a horizontal position in the center of opposite sides of the 4 by 4-foot hover or in the center of the ends of the 4 by 6-foot hover so that the center of the porcelain lamp socket is 3 inches above the bottom edge of the hover (fig. 4).

The following materials are needed for a 4 by 4-foot brooder:
- One piece of 4 by 8-foot, ¼-inch plywood or 1/8-inch pressed wood (to be cut into one 4 by 4-foot top and four 1 by 4-foot sides)
- Four cleats 1 inch by 1 inch, 4 feet long, to which the top and sides are nailed.
- Four pieces of 1 ½ by 1 ½-inch lumber, 16 inches long, for corner posts or legs
- Two porcelain electric lamp bulb sockets (Porcelain lamp sockets are necessary for these lamps)
- One 150-watt, 115- to 120-volt projector or reflector Mazda spot or flood lamp and one 250-watt R-40 Bulb Drying lamp
- Twenty feet of rubber-covered electric appliance cord with plug and cap[*]

No special provision need be made for ventilation. That which takes place through the open space between the lower edge of the hover and the floor will be ample. As the chicks or poults grow larger and need more air and less heat, bricks or blocks can be placed under the legs to raise the hover 2, 4, or 6 inches higher. When feed and water are to be placed under the hover, or the floor litter is to be removed, one side can be raised to the desired

---

[*] The simplest and cheapest way to make a power cord for the brooder is to buy a two-wire extension cord and cut off the socket end.

Figure 4. Inside of a 4 by 4-foot hover equipped with two lamps.

height and held in place by a hook suspended from the ceiling of the brooder house.

This type of brooder with the abundance of light within makes it convenient to feed and water the chicks or poults under the hover during the first day or two; after that, the feed and water can be moved outside. The abundance of light beneath the hover and the feeding of baby turkeys under the hover during the first few days have proved especially advantageous for starting poults.

No thermostatic regulation of the temperature is needed, since the chicks readily adapt themselves to their own temperature requirements and comfort in a brooder of this kind. Whenever it is observed that a considerable number of the chicks find it comfortable at the edge of, or outside, the brooder, the hover should be raised 2 to 4 inches to admit more air and to lower the temperature beneath it. If two lamps are in use, one can be turned off.

The curtains used at the Experiment Station when needed to prevent floor drafts or to conserve heat under the hover during cold weather are strips of cloth 8 inches wide and 4 feet long made from feed bags.[*] The strips are attached to the sides of the hover

Figure 5. Chicks under the brooder at night.

with thumbtacks so that the bottom of the curtain is ½ to 1 inch above the floor litter. The bottom of the curtain should be hemmed but need not be slit. When the hover was used in a room provided with another source of heat so the temperature seldom went below 40 °F, curtains were not needed unless there was evidence of a floor draft which caused the chicks to congregate at one side of the hover. When that occurred, a curtain was attached to the exposed side or sides opposite those where the chicks congregated. A curtain on one or two of the exposed sides gave effective protection against floor drafts. When day-old chicks were started in an uninsulated colony house during cold weather (10 to 20 °F) in January 1941, it was necessary to use curtains on three sides of the hover during the first week to conserve the heat under the hover. Afterwards, two of the curtains were removed; one was left to prevent floor drafts. Also, a corrugated cardboard band 12 inches wide was used to keep the chicks within 1 to 2 feet of the

* Presumably burlap sacks.

hover during the first few days. Feed and water were provided under the hover during the first 2 or 3 days.

In usual practice under average brooding conditions during April or May or in a room where supplementary heat is provided, the 250-watt lamp would generally be used during the first week or 10 days of the average 6-week brooding time, when the chicks or poults need the most heat.[*] After that time, the 250-watt lamp would generally be discontinued, and the 150-watt lamp used for the rest of the brooding period. On this basis, the cost of operating the 250-watt lamp 10 days (at 18 cents a day of 24 hours with electric current at 3 cents a kwh.) would be $1.80, and that for the 150-watt lamp (at 10.8 cents a day for 32 days), $3.45. The total cost of electricity during a 6-week brooding period would, then, be $5.25.[†] In warm weather, the cost of the electricity would be lower, since the brooder lamp would either be turned off, or one of the brooder lamps replaced by an ordinary 15-, 25-, or 50-watt Mazda light bulb to provide an attraction light and a small amount of heat during warm days or nights. Likewise, the small Mazda bulbs could be used during the latter part of the brooding period, when an attraction light and only a little heat are needed. On the other hand, brooding during cold weather in a cold room with both lamps in use much of the time would cost correspondingly more, just as the cost of brooding during cold weather is greater regardless of the source of heat.

The effective insulation against heat loss which this type of hover provides can, however, be expected to prove economical in the use of electricity regardless of the kind of electrical heating element employed. In two of the tests, meter readings were made to secure the electric current requirement of the lamp brooder in comparison with a conventional brooder equipped with thermostatic heat regulation, fan, and special ventilation.[‡] The first test

---

[*] I prefer to use two bulbs at all times, in case one burns out. Thus, I would use two 125-watt heat lamps instead of one 250-watt lamp.

[†] This works out to $10.50 at 6 cents/kwh, which is what I pay.

was conducted in uninsulated colony brooder houses during January and February, 1941, and the second, in adjoining brooder pens during April and May. In both cases, the electric current consumption was somewhat less for the lamp brooder.

The principal advantages of the electric brooder described are simplicity, low first cost, and effective insulation at practically no cost. Some poultrymen may be inclined to add needless complications and expense, such as thermostatic heat regulation, special ventilation, or other nonessential items or gadgets which would tend to offset the primary advantages and purpose of this type of brooder. The contention that such additions are needless is based upon the results of a year of almost continuous use of five of the brooders at the Station's Poultry Plant. Hundreds of chicks have been brooded at all times of the year under widely varying conditions. In all instances, satisfactory results were secured with these simple, inexpensive brooders. At no time was there noticeable evidence of a need for thermostatic heat regulation, additional ventilation, or other items that would make these brooders more complicated and expensive.

---

‡ Fancy electric brooders with thermostatic heat control, circulating fans, and heavy insulation were common in the Thirties and Forties, but have not been manufactured for decades.

Chapter 8. More About Insulated Heat Lamp Brooders

The overhead heat-lamp brooder is simple and effective, but it uses a lot of electricity, and is extremely dependent on the draft guard. The Ohio brooder uses less electricity and is less dependent on the draft guard, especially if you use curtains on the draftiest side.

The brooders are easy to build, even if your carpentry skills are as rough as mine. I timed myself when I built my third brooder, and it took precisely two hours. (It probably took three hours the first time.)

8.1. Brooder Capacity

A rule of thumb for the capacity of these brooders is 12.5 chicks per square foot of hover area and 2.5 watts per chick during cold-weather brooding. Thus, a 4x4' brooder, with 16 square feet of area, can house 200 chicks and calls for 500 watts of light for cold-weather brooding, where the brooder house is subjected to freezing temperatures Brooding with overhead heat lamps would take 1,000 watts. Also, the wattage can be cut in half quickly, often after the first week, which is not something I'd do with overhead heat lamps.

Size	Area	Initial Wattage (First Week)		Chicks
		Sub-Freezing Weather	Above-Freezing Weather	
2x2'	4 sq. ft.	125W	75W	50
2x4'	8 sq. ft.	250W	125W	100
3x3'	9 sq. ft.	250W	125W	112
4x4'	16 sq. ft.	500W	250W	200

Figure 1. Capacity of insulated heat-lamp brooders, assuming a six-week brooding period (capacity can be doubled, with some risk, for a 2-3 week brooding period)

8.2. Construction

The brooder should be built of the lightest plywood available. Quarter-inch plywood is thick enough. (I made my first brooder out of 1x12 boards, with a masonite top, and it is very heavy and awkward.) The four side pieces are nailed to the legs so the legs stick down 4 in. below the sides. Four pieces of 1x1 in. or 1x2 in. lumber are nailed along the sides to make a rim that the lid will be attached to. The lid is nailed onto this rim. A brace across the middle of the lid keeps it from sagging.

Because the sides extend for four inches above the lid, you can pile four inches of shavings onto the top of the brooder box. This gives an insulation value of about R-20 at practically no expense. As the chicks get older, some of them will begin to roost up on top, where they enjoy the warmth coming up through the shavings. This double-decker effect reduces the crowding that might otherwise occur as the chicks get larger.[*]

Caulking the seams helps the brooder to retain heat, but is not absolutely necessary. Painting the underside of the lid with white or silver paint to reflect more light downward is an extra touch that will increase the amount of heat reflected directly onto the chicks. Some farmers used sheet metal for this purpose. Don't use foil, since the chicks will peck at it and it will soon be hanging down in strips.

The lamp sockets are centered vertically between the lid and the bottom of the side pieces (which is four inches from the bottom of the side pieces). Use electrical boxes (preferably plastic boxes, which can't present a shock hazard to the chicks even if a wire comes loose) and porcelain lamp sockets. I tried screwing porcelain light sockets directly to wood, but the wood sometimes became scorched, so I gave it up as a bad idea. The electrical boxes will separate the hot wires and sockets from the wood.

---

[*] Ordinary brooders can accommodate only 10 chicks per square foot during the latter part of the brooding period, because the chicks are too big to fit under the hover.

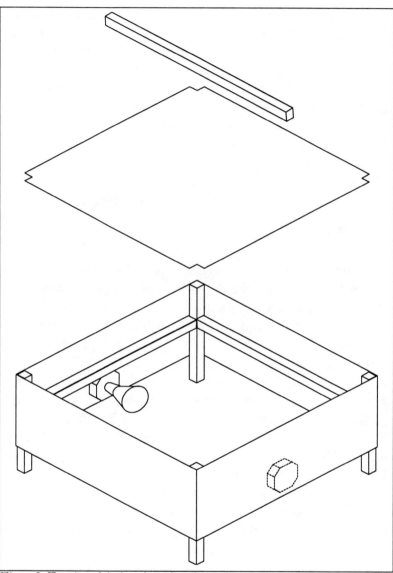

Figure 2. Framing of the heat-lamp brooder

I like to attach the outlet box to the brooder first (with bolts), then drill a hole through the brooder to the middle of the box, and install my power cord through the hole. I knot the cord as a strain relief, leaving plenty of length to attach the wires to the lamp socket.

I always buy cheap two-wire extension cords, rather than buying lamp cord and plugs separately and wiring them together. The molded plugs on extension cords are more weatherproof than hand-wired plugs, and extension cords are cheaper than buying the pieces separately. I snip off and throw away the socket end of the extension cord, and wire the rest to the lamp sockets.

It's simpler to have separate power cords, one for each lamp, than to wire the two sockets to the same cord. Also, when it comes time to reduce lamp wattage, it's less disruptive to the chicks to unplug one lamp and leave the other burning. You can come back later, after the lamp has cooled, and replace it.

8.3. Using the Brooder

This kind of brooder works well without a thermostat. Because it is open on the bottom, the chicks can move out beyond the edge of the hover if it gets too hot inside, and they can crowd close to the lamps if they get cold. I have always found thermostats to be frustrating because they're hard to set properly. Experiments made by the Ohio Experiment Station found that this kind of brooder used less electricity than contemporary insulated commercial units with thermostats. (Nobody makes such units these days, anyway.)

Insulated heat-lamp brooders are less sensitive to floor drafts than overhead bulbs, because the brooder itself provides a little bit of draft shielding. Floor drafts can still be a problem, though. You should still use a draft shield during the first week.

One way to deal with drafts is to put a curtain on one side, facing the strongest floor draft. This can be a simple strip of cloth, held on with thumb tacks or staples, that extends to within an inch of the floor.

You can also extend two sides all the way to the floor, giving two solid sides and two open sides. The two solid sides need to be on opposite sides, since you don't want to form a corner in which the chicks could pile and suffocate. I find this modification to be very convenient for winter brooding, especially in my drafty brooder houses, which were designed with adult hens in mind, not chicks.

One advantage of having two sides that extend to the floor is that the brooder won't sink into loose litter the way it does when supported solely by its 2x2 legs.

As the chicks get older, you can raise the brooder on bricks or pieces of scrap lumber to give them more ventilation, and you can reduce the wattage of the brooder by replacing the bulbs with smaller ones. I start with 125W heat lamps and switch to 65W indoor floodlamps later on. The time to switch lamps or raise the brooder is when you notice that the chicks are sleeping outside the edge of the brooder in the cold of late night or early morning. This means they need less heat.

If the weather in the house is likely to drop below freezing when you start a new batch of chicks, you should put the feed entirely underneath the hover for the first three days, with the quart-jar waterers just outside the hover, with the jars touching the side wall so the base extends a little way into the house. This will encourage drinking and will keep the water from freezing. You can also put a quart-jar waterer entirely inside the brooder, but raise it up extra high (on a scrap of 2x4, for instance), to ensure that no chicks get soaked in it.

I have had excellent results with these brooders at all times of year. My experience is that the chicks seem happier in my drafty houses with these brooders than with overhead heat lamps, especially in very cold weather. And the reduction in electricity usage (which is cut by two-thirds) is very welcome.

Chapter 9. Battery Brooders

Figure 1. Fashions change, but battery brooders remain the same.

Battery brooders don't run on batteries! They're called this because they were used in rooms with whole arrays—batteries— of identical brooders. A battery brooder is a brooder with a wire floor and a manure pan, a heating element, and built-in feed and water troughs.

Battery brooders are stackable. The photo above shows a battery of six identical brooders sitting on a frame with casters. It is easy to take care of a great many chicks in a small space with these brooders.

Battery brooders have a number of advantages over conventional floor brooding. They provide the chicks with an environment where it is impossible for them to get chilled or lost. They learn where to find feed and water very quickly. The wire floor prevents them from being exposed to manure and litter-borne diseases like coccidiosis and roundworms.

9.1. When to Use Battery Brooders

Battery brooders are particularly useful if:
- You are raising birds that are unusually prone to mortality early in the brooding period (such as turkeys and gamebirds).
- It is too cold to use your normal brooder house for newly hatched chicks.
- You brood several groups of chicks of different ages at the same time.
- You sometimes need temporary or overflow brooder space.
- The weather is warm enough (or your grower house is snug enough) that the chicks don't need brooder heat for more than a couple of weeks.

My experience with battery brooders is that they provide a significant reduction in first-week mortality. This is especially true of turkey poults, but is true of chicks as well.

Many farmers have worked out systems where they brood chicks in battery brooders for 3-14 days and then move them out to a conventional brooder house. This uses the battery brooders only when the chicks are very small and vulnerable to cold, and shifts them to a regular brooder house when they need more space.

This can increase your chick capacity, since a new group of chicks can be installed in the battery brooders as soon as the old one is transferred to the brooder house.

You can keep chicks in batteries throughout the brooding period if you want, but they need a lot more space as they grow, and the amount of manure produced by older chicks makes brooding them in batteries fairly unpleasant. When the chicks are small, the amount of manure is small and cleaning the dropping pans is not very unpleasant. This changes as the chicks get older.

I am happy with chicks in my basement until they're about a week old.

9.2. Where to Find Battery Brooders

As with so many things in the poultry industry, battery brooders have been around practically forever, but went out of fashion when the vertical integration of the poultry industry reduced the number of farms and narrowed the methods being used. The definitive book on the subject was written in 1930[*].

New battery brooders are still being manufactured by GQF Manufacturing in Savannah, Georgia (http://www.gqfmfg.com) and at least one other company.

Old units exist in large numbers and often sell for low prices. Some of the more ancient brooders may be hot-water units or have no provision for heating at all, but were meant to be used in a room heated to brooder temperature. These can be retrofitted to thermostatically controlled electric heat. GQF sells an appropriate heating element/thermostat combination.

Electric brooders made before the advent of three-wire power cords (with a grounding prong) should be given a new power cord that connects the metal shell to ground. Otherwise they're perfectly modern.

9.3. Where to Put Your Battery Brooders

Like most electric brooders, battery brooders are uninsulated and can't be used in unheated rooms. The manufacturers usually specify 70 °F. If you have several brooders in a small room, the heat

---

[*] Arndt, *Battery Brooding*, Orange Judd Press, 1930

Figure 2. Feeding time

from all the brooders may keep the room warm. Otherwise you will need supplemental heating.

I have three battery brooders in my basement, which never falls below 50 °F in the winter. This temperature is really too low for the battery brooders, but I get good results if I drape an old sleeping bag over them. However, I use an unusual type of brooder that has openings only in the front. This provides inadequate amounts of feeder and waterer space for older chicks, but it makes it easier to keep them warm. Most brooders are open on three sides. In any event, it's far better to put the battery brooders in a warm room, since fussing with insulation and blankets is time-consuming and worrisome.

If you have a stack of brooders, instead of just one, the brooder on the bottom of the stack will be harder to keep warm than the others. One trick you can use when the brooder room is cooler than it ought to be is to leave the bottom brooder on, but to put no chicks in it. This will help heat the brooders above.

Putting your brooders anywhere in your home will work fine for a couple of days, but is a big mistake if you're going to keep the birds for more than a week. Battery brooders smell, and the dust from the chicks (a combination of down, dandruff, and manure) coats everything. Also, spilled feed from the troughs attracts rodents. Lots of people use battery brooders in their houses—I do—but it has serious disadvantages that are best dealt with by moving them to an outbuilding. By partitioning off a portion of a barn, shed, or garage and insulating it, it will be possible to keep it warm without too much heating expense.

The room needs to be big enough that the brooder can be accessed from at least three sides without moving it. There has to be enough room in front to slide out the dropping pan. For a standard 30 in.x36 in. brooder, there needs to be clear space on the sides about two feet wide, and on the front about four feet wide. Thus, a single stack of brooders needs a space roughly seven feet square.

If you have multiple stacks of brooders, they can share the same aisle space, and they use space more efficiently. Stacks can also be placed back to back, since access to the back is not absolutely necessary (though it helps). In the old days, many people converted the traditional 10'x12' portable brooder houses to hold two stacks of brooders, and this left plenty of room for feed and equipment.

Ventilation becomes very important after the first week or so, when the moisture from the chicks' respiration and manure becomes significant, and their manure production bothersome. Windows are probably adequate sources of ventilation for a few stacks of battery brooders. The windows shouldn't direct a blast of air directly over the brooders. Double-hung windows would be better than sliding windows, since you can get quite a bit of ventilation by opening the top sash only, and most of the air movement will be above the brooders. In hot weather, you'd open both top and bottom sashes.

Windows increase the level of light indoors, and battery brooder chicks are notorious for their tendency to feather-picking and cannibalism, which is also encouraged by high light levels. High light levels also tend to make the chicks more nervous and panicky than otherwise. Many old-time poultrymen sprayed their window panes with red lacquer, since chickens don't react much to red light, but it gives enough light to work by. (This is not to say that chicks should be kept in the dark, because they need light to eat and drink. But bright light encourages chicks to a level of activity that the battery environment cannot support.)

Really big installations need forced-air ventilation. This is the same ventilation problem faced by ordinary confinement broiler growers, and fans and intelligent fan controllers can be purchased from vendors catering to such businesses, such as FarmTek (http://www.farmtek.com).

Fly problems can be avoided by screening all openings, putting an automatic closer on the door, and removing manure frequently. By cleaning the dropping pans every day or two and removing the manure from the building, fly eggs won't have time to turn into a new generation of flies inside the building.

Rodent control should be through having a tight floor, a good fit on doors and windows, and some kind of trapping or baiting program. Feed and trash containers should be rat-proof (galvanized trash cans are good). The area around the brooder house should be kept clear of weeds and clutter, since these provide nesting places for pests.

9.4. Setting Up the Brooder

Before the chicks arrive, the brooder needs to be set up and warmed up. This provides no difficulty except for setting the thermostat and perhaps adjusting the gates on the sides so the chicks can't get out into the feed and water troughs.

9.4.1. Setting the Thermostat

The unit will have a brass thermostat wafer mounted on an adjusting screw to a bracket. Beyond the thermostat is a micro-switch that turns the brooder on and off. The disk gets thicker as it gets hotter, eventually pressing against the switch and turning off the brooder heat. When the wafer cools slightly, it shrinks away from the switch, and the brooder turns back on.

The switch has an audible click which makes setting the thermostat easier.

To set the thermostat, you need a thermometer that can be read without opening the brooder. Indoor/outdoor thermometers and the probe thermometers sold for incubator use are good for this. A practical place to set the thermometer is fairly high up in the back wall, near the thermostat. This will give a higher reading than the temperature at chick height, but it's a spot where the thermometer can be used at any time. Once the chicks arrive, a thermometer at chick height will become covered with manure, or will have chicks perching on it, and its readings will be worthless.

If the unit has a socket for an attraction light, screw in a 7 ½ watt bulb. This will probably switch on and off with the brooder, which makes adjustment a little easier.

Back the thermostat adjusting screw until the thermostat wafer is as far from the switch as possible. Turn the brooder on. The temperature will rise. Go away for an hour or two. When you come back, the brooder should be as hot as it's ever going to get. If this doesn't read 100 °F or more on your thermometer, you don't have enough heat. You may need to consider restricting the airflow in the brooder. You can use saran wrap as a draft guard by draping it between the side of the brooder and the troughs. Some kind of insulation on the top will help, too. Use a higher-wattage attraction lights (for example, 100W) if all else fails.

Once you've demonstrated to your satisfaction that the brooder has enough power to get too hot, turn the thermostat adjusting screw until the switch turns off, and then turn it some more.

The brooder will now begin to cool down, and within a few minutes it should turn back on again.

I like to get the temperature down to about 100 °F-110 °F at chick level right under the heater, on the grounds that every brooder needs an area with extra heat where the chicks can go to warm up quickly if they feel chilly. The edges of the brooder will be much cooler than the region right under the heater.

Don't be sure you've got the temperature set right until you've gone away for an hour or so and then checked it again. Sometimes it takes a while to settle.

My experience is that setting thermostats to a precise temperature is very frustrating, but precision isn't as important in brooders as with incubators, and an error of a couple of degrees won't hurt anything.

For consistency in measurement, always use the temperature reading at the moment where the brooder turns off as the "real" temperature. The temperature varies by several degrees during a cycle, and you can end up chasing your tail if you don't always use the same procedure for deciding what the "real" temperature is.

If the temperature is too high, advance the screw to move the wafer closer to the switch. If the temperature is too low, back the screw off.

If the temperature is too variable, the wafer may be failing. It's good to keep a spare. The wafers unscrew easily from the adjusting bolt, and you can test all your wafers at once by running a stream of warm water in the sink and passing a handful of wafers through the stream. The good ones will expand quickly, while poor ones will expand slowly. Bad ones won't expand at all.

It is also useful to have a spare switch. You can buy thermostat wafers, switches, and thermostat/switch assemblies (including the adjusting screw and the mounting bracket) from GQF and others.

9.5. Chick Capacity for Battery Brooders

The chicks start to outgrow a battery brooder very quickly. I've put some recommendations into figure 3. As far as I can tell, none of today's broiler suppliers make recommendations for battery brooding, so I have estimated the space needed by modern broilers by assuming that, for example, an eight-ounce broiler chick needs the same amount of space as an eight-ounce standard-bred chick. Broilers have such an astonishing growth rate that the two sets of figures in the table seem to be for different species.

Crowding causes all sorts of problems, including uneven growth due to competition for the feeders and waterers, higher mortality, and a terrific increase in unpleasant smells. The numbers in the table are *not* conservative, and you exceed them at your peril.

Chicks per 36x30 in. Brooder	Brooder Space per Chick		Age of Chick, Weeks	
	sq. in.	sq. cm.	Standard Breeds	Modern Broilers
120	10	60	0-1 weeks	-
60	18	115	1-3 weeks	0-1 weeks
36	30	195	3-6 weeks	1-2 weeks
18	60	385	6-12 weeks	2-4 weeks
8	135	870	12+ weeks	4+ weeks

Note: Floor space requirements for standard breeds are taken from Hoffmann and Gwin's *Successful Broiler Growing*, p. 210. I derived the values for modern broilers by giving the same floor space to birds of the same weight. As you can see from the table, modern broilers can outgrow a battery brooder very quickly.

Figure 3. Battery brooder space per chick

9.6. Grower Pens

If the chicks are going to be kept in battery brooders for more than a few weeks, they are typically transferred to grower pens, which are unheated pens that are taller and better ventilated than battery brooders, and generally have a sturdier, wider-mesh floor (for instance, 1x1 in. welded wire instead of 0.25x0.25 in. hardware cloth). In the heyday of battery brooders, broilers were brought to market age in these cages. This was in the days when a broiler of between 2 and 2 ¼ pounds live weight—about the size of a so-called "Cornish Game Hen"—was considered to be of market age.

Growing broilers to larger sizes is not very satisfactory, because of problems with breast blisters, the increasing amount of space required per bird as they grow, and the immense amounts of manure generated by larger broilers.

9.7. Using Battery Brooders

Arndt covers the basics of using battery brooders so well that I am going to turn the rest of the chapter over to him (this text is excerpted from *Battery Brooding*, pages 169-192):

AMOUNT OF ARTIFICIAL LIGHT

When the day old chicks are placed in the batteries they are in new surroundings and do not seem to know what it is all about. The wire bottom bothers them at first and they seem to be lost. Many systems have been attempted to acquaint them quickly with their surroundings so they will get busy at the feed and water pans as soon as possible. By putting fifty or sixty watt light bulbs in all the sockets in the nursery room a bright light is thrown on the feed and water pans, but at the same time the centers of the battery trays are dark.

The day old chicks are attracted to this light. They seem to have that habit in common with moths that fly around the street lights at night, so they endeavor to get as close to the light as possible. This naturally brings them almost in direct contact with the feed and water, so only a few hours usually elapse before a good many birds are eating and drinking. As soon as they have discovered the water and the feed and know where to find it, they appear to become comfortable; so it is a pleasure to watch them running from one part of the battery to another.

Operators not prepared with the bright lights when the chicks arrive are forced to put shallow pans or papers on the wire floor so as to acquaint the chicks with their feed and get them started eating before they get too hungry. Handling the chicks or paying any attention whatever to them directly after they are placed in the batteries, is a waste of time and labor. Much better success will follow then the bright lights are provided.

Only one factor may prevent them from coming to the light within a few hours; a temperature not high enough for their comfort. When chicks are not warm enough they will not respond. Of course, it is possible that poorly hatched chicks, chicks of low vitality or very hungry ones will not respond to the light treatment.

These bright lights are kept on for twelve hours—from seven o'clock in the morning until seven o'clock at night is the most convenient period for the average operator.

If the lights are equipped with a time switch, it will not be necessary for the operator to be present to turn the lamps on or off. It is an advantage to keep the bright lamps in their sockets for three or four days or even the first week, but then to change them to ten or fifteen watt bulbs. The birds no longer need the bright lights so unless these are removed shortly, especially with the White Leghorn chicks, the intense light is sure to do more harm than good. Besides, it is time to train the birds to live, eat and drink in as dim a light as possible for the operator to work in. The dim lights, possibly not more than fifteen watts in any one socket, should be used at all times in the finishing stages of brooding. Bright light for the older birds seems to accomplish nothing more than to make them irritable, panicky, and to get them into bad habits such as feather pulling and cannibalism.

A large number of our cases of cannibalism in batteries are caused by too much light in the room. This is another reason why all natural light and sunlight have been excluded, because a ray of sunshine in a battery tray is almost sure to start a lot of cannibalism and feather pulling. In all three battery rooms after the first week in the nursery room dim lights are used twelve hours a day throughout the entire brooding period.

When battery brooders first came to be used, it was advocated that the birds be allowed to have access to the light twenty-four hours a day, but experience has proved this to be costly in more ways than one.

In one experiment the birds consumed considerably more feed when allowed twenty-four hours of light a day. They were much lower in vitality at the end of a six week period and had made less growth than the birds given twelve hours light. In an experiment with eighteen hours of light the birds at the end of six weeks had not made better growth, weighed no more than bird that had received twelve hours' light, though they seemed to be in better physical condition than the birds that had received twenty-four hours' light.

These experiments show that there is no reason why money should be spend for electricity six to twelve hours each day and not have anything to show for it.

Further experiments with small lots of birds proved that the best weight and vitality possible could be assured by keeping the lights on for six hours and then off for six hours and so continuing.

FILLING THE FEED PANS

The feed pans and the amount of feed they contain at different stages of battery brooding is important and should be carefully watched....

When baby chicks are brought into the nursery room, any little thing that can be done to help them start off quickly is of great advantage. One thing that has been of big help is to have the feed pans full to overflowing when the chicks first arrive. When their little heads come out of the battery to see what the bright light is all about, the feed should be as close to them as possible. This will not be the case if the pans are only partly full. If they are filled completely for the chicks' first feeding, it is not necessary to fill them again for two or three days, as only a small amount of feed is eaten by such young birds. In fact, sometimes the pans may go for the entire first week before more feed need be brought in.* When the birds learn how to eat, they will reach down in the pans for the feed so that before they learn the pernicious and wasteful habit of throwing feed, it will be too low.

While it is wise to be extremely liberal with the feed for the first feeding, it is not advisable to maintain this liberality at any time after the first week. From this time forward it is better to keep the feed pans only about half full for the rest of the brooding period. By doing this and allowing the birds to reach the bottom

---

* Andt wrote this before the development of modern broilers and their incredible appetites. The troughs for young broiler chicks may have to be topped off more than once a day.

of the feed pans at least once a day, the operator may be sure that no old feed is left in the pans from day to day, but that all is being cleaned up completely, not being sorted over and pulled out of the pans as is sometimes the case.

When the birds are hungry, they will eat and feed as fast as they can get it, but when their appetites have been appeased, they will sort it over with their beaks and throw more or less of it on the floor in their compartment, thus wasting it.

If the feed pans are properly constructed, higher on the outside than on the inside edge, and have guard incorporated on the inside to prevent the birds from beaking the feed into their compartments, little feed will be wasted. Many battery brooder operators lose considerable feed because they fill the feed pans too full.

By keeping the floor clean, the small amount of feed sometimes thrown onto the floor may be easily swept up and used to finish up some older birds for market.

WARM WATER AT FIRST

If the day old chicks are to be received tomorrow, it is well and advisable to fill the water pans today. By doing this, part of the water will evaporate and increase the humidity in the room, and the balance will be warmed almost to the temperature of the room. It is highly desirable that the first water the chicks receive should have the chill removed from it. It is an advantage to have all the water they receive for the first two days near nursery room temperature. The birds drink more of it at that temperature and there is less chance for it to cause them trouble than if they are forced to drink cold water.

Water in battery brooding is of exceeding importance and should be kept before the chicks at all times. Many operators are careless on this point. They allow the water pans to become empty and remain so for hours at a time. Almost invariably birds handled in this manner fail to produce economic growth.

If water is piped into the building and spigots are located in each room, it will save time in handling the birds. A large size sprinkling can from which the spreader end has been removed is a convenient thing with which to fill the water pans.

A still more convenient method is a length of hose attached to the spigot with a shut-off on the free end. With such an attachment it is possible to move quietly from one battery to another and water the entire room quickly and easily....

The water pans should be cleaned frequently as the birds when drinking deposit feed from their beaks in the water. If left for long in the pens the warmth of the room soon starts this feed to ferment. A scum forms on the pans and the water becomes polluted. The pans should be removed and scrubbed with a stiff brush at least three times a week. It is advisable to use a strong disinfectant in the scrubbing water so as to kill any fungus growth that may have started.

ADJUSTMENT OF BATTERY SIDES

Almost every battery manufactured today is equipped with some kind of adjustment so the chicks cannot get out of the compartments. The adjustments must be made accurately and they should be easy to control and to change readily as occasion demands.

If the battery cannot be closely adjusted, especially during the first week, a large number of chicks will be on the floor every morning. This is not pleasant because it is an exasperating task to crawl around on the floor to capture the little fellows. It is all the more unfortunate where several lots of chicks are in the same room; for there is usually no way of telling in which compartment they belong. It is something to be very careful about when purchasing a battery. To let a day old chick through does not require a much larger opening than for the same chick's head, so the battery must be adjusted carefully, otherwise the larger chicks in each tray will not be able to get their heads out to feed and drink, though the smaller ones will squirm through and fall to the floor. The adjustments on the battery should be so flexible that they can be changed at any time on short notice by simply moving them up and down, or sliding them sideways—depending on the type of battery....

By adjusting the battery carefully and closely the first day after chicks are put in and filling the feed trough level full (smoothing the surface with one's finger) it will be possible to note the following day just where the chicks can reach through and eat, and where the adjustment is too close. The chicks in eating from the full tray make little "V"-shaped ditches in the feed where they can get their heads through. By going over the batteries carefully the second day, it will be easy to locate the spots where the adjustment is too close as the feed in these spots will not be disturbed.

When Chicks Crowd or Are Uncomfortable

Chicks will show by their actions especially during the first two or three weeks just how they feel. If they make that peculiar "cheeping" cry and stand all huddled together in the center of the battery, where the temperature is a few degrees warmer than at the outside, they are cold and uncomfortable. They will not eat when cold and therefore will not grow. So when this sound is heard from many chicks at a time the operator must increase the heat and check up on the thermometers.

Chicks begin to cry, stand around in little bunches and slowly work their way to the center of the tray as soon as the temperature drops a few degrees more than it should. Only by reducing the temperature gradually-one degree a day-can the tremendous drop from 90 down to 70 degrees in a period of three weeks be accomplished without discomfort to the chicks.

On the other hand, it is also easy to determine when chicks are too warm. Then a large percentage of them will stand near the outside edges of the battery panting, and with their mouths open. If they are extremely warm, besides panting, their wings will be hanging loosely at their sides. This condition is as bad as having them too cold. In this condition they will neither eat nor thrive.

These conditions of being too hot or too cold are responsible for the poor start that some battery chicks make. They indicate the necessity for accurately controlled temperature. Only by this means can we give them the temperature they require and make them feel happy and contented in their surroundings.

When one has had experience he will readily realize when they are feeling good as they will be eating, running about, and singing contented little songs.

After the first operation or after the first few weeks of operating, one will readily recognize the moment the door is opened whether the chicks are contented or whether something is wrong as they make the two distinctly different sounds mentioned.

By having an accurate control of the Oil Heaters or other heat supply, one will be able to secure and hold the maximum vitality of the chicks from the time they are a day old until ready for the other stages of brooding.

The chicks in the other two sections, especially those older than four weeks, are not so susceptible to changes in temperature; in fact, they will really do better if the temperature fluctuates two or three degrees up and down the scale. The older birds will also quickly show the effect of too much heat in their room in the same manner as the little fellows. When they are overheated they will pant and also refuse to eat. Instead of gaining, too warm a temperature will undoubtedly cause them to lose weight. This condition of overheating must be remedied quickly if the older birds are to continue growing properly. This can be accomplished usually by increasing the amount of fresh air coming into the room, especially during the warm months. The automatic control of the exhaust fans will be a big help in guarding against this condition.

THE NUMBER OF CHICKS TO A TRAY

Though this has been discussed in previous chapters, it should be presented in more detail here.

When using this type of battery in connection with the diagram already shown it is necessary to know its capacity. We are interested now less in the outside dimensions than the actual floor space available to the birds. Some manufacturers advertise the size of batteries at maximum dimensions, that is, including the space occupied on the outside by the feed and water pans. These dimensions are worthless to the chicks. We are interested only in the space in which they are to be confined. The compartments in the average battery of this type usually have from five hundred and fifty to six hundred square inches of floor space.

This is the convenient size that can be handled with ease by the average operator. Usually these small compartments are from fifteen to eighteen inches deep and from thirty-two to thirty-six inches wide. This type of compartment means that the operator will not have to crawl half into the battery in order to reach and get a hold of the chicks when they run away from him. In the average battery formerly manufactured and widely sold, unless the operator was very tall and had an exceedingly long arm reach it was very difficult to reach the chicks when they were frightened. In the large compartments when the door was opened the chicks had a habit of rushing back away from the operator and piling up on top of each other in the corners. They very often mutilated each other by scratching deep gashes in each other's backs with their sharp toe nails. These troubles are entirely eliminated in the shallow compartments.

In the small size compartments outlined above it is practical to place forty day old chicks. These forty day old chicks will not outgrow this compartment until they are about four weeks of age at which time it is necessary to transfer them to the finishing room.

If we use the same size compartment in the finishing room our equipment is uniform throughout. When the forty chicks in the one compartment in the nursery room are transferred only twelve of them are placed in each compartment in the finishing room. They seem almost lost in this compartment at this age but this is one of the important times in a chick's life and it will pay the operator to allow them plenty of room to grow and develop. The more room they have at this age the faster they will grow and the more complete will be their feather growth. Another tremendous advantage of this system is the fact that these twelve chicks will continue to grow and mature without further handling or disturbance until they are ready for market. This you can readily understand eliminates a great deal of labor which would otherwise be necessary if more than the twelve chicks were placed together when they were four weeks of age.

As the chicks near market age during the eighth week, the largest ones will be removed for market. This will allow more floor space for each chick left. It is surprising how these remaining birds will grow during the week following the removal of the largest ones. This seems to prove that if we could afford to allow more space to each bird, we could obtain still better growth. It may be possible that the older ones were more or less intimidating the smaller ones preventing or checking development of these smaller birds.

CLEANING BATTERIES

The opportunity to clean our trays and batteries in this system of brooding is when birds are shifted from one room to the next. After birds have used a battery for three weeks, dust and dirt from the circulation of air has accumulated. Small quantities of droppings will have been deposited in the crevices so it is time to wash and thoroughly disinfect every part of each battery.

If the wire bottoms are removable as they should be, they can all be taken to a big water-tight box constructed for soaking and cleaning them. Attempts to clean these bottoms in a few moments are useless wastes of time because the particles resist this type of cleaning. It is always an advantage to remove them from the room and soak them for five or six hours or over night in a big wooden vat. After they have soaked it is easy to scrub them off with a broom. They should then be set in the sun to dry.

If water pressure is available in the battery room the balance of the battery may be cleaned thoroughly with a hose. This clean-up every three weeks will also give an opportunity to wet down the floor thoroughly and give it a good scrubbing.

If the inside walls and ceiling are built of materials that will stand water, it will be a great advantage to use the hose freely and forcibly to remove dust and cobwebs. By scrubbing all the batteries and the entire room, the possibility of carrying over disease germs or bacteria from one brood of chicks to another is greatly reduced if not wholly prevented. This clean-up every three weeks will keep the room at all times in sanitary condition. If wooden walls and insulated ceilings have been painted or are water proof the work at this time will be greatly facilitated.

If the concrete floor has been carefully laid with a slight fall to each of the intake ducts, almost all the water will pass out of the room to the fresh air openings which can then be flushed out.

If it is thought that there still may be disease germs present it is a good plan to use a treatment of formaldehyde gas such as used in disinfecting homes after contagious diseases. By closing up all of the openings, starting the circulating air shaft and freeing this gas it will only be a few hours before all disease is destroyed. The room can be freed from these fumes in a few minutes by opening the ventilators and the intakes and allowing fresh air to come in.

CLEANING DROPPING PANS

The droppings may be removed by one or the other of two satisfactory ways.

A more expensive way is to use a heavy, tough grade of paper on the dropping pans and to remove both paper and droppings quickly every morning without removing the dropping pans. This system eliminates the handling of the pan each day, but in a large operation the purchase of a paper of a grade good enough to withstand the moisture of the droppings is expensive. In case there is a sale for the manure the paper might prevent sales.

In the nursery room where the droppings are small and fairly well dried out and in which the quantity is small, old newspapers may be used in place of the expensive moisture-resistant paper. This answers for only a short time, because as soon as the moisture accumulates in the droppings the newspapers becomes soggy and tears easily when attempts are made to remove it.

The other system of cleaning is to remove the dropping pans and scrape them every morning. This is more practical than using paper but it is a hard, dirty job. Cleaning dropping pans is probably the most expensive one operation in battery brooding. Even the clever fast operator finds it difficult to handle more than one pan a minute. In scraping dropping pans it is well to have a wheelbarrow or a carrier in the aisle so that the end of the pen can rest on it while the operator can scrape with a steel scraper about one foot wide.

After the pans are scraped, many operators find it an advantage to sprinkle them with finely ground peat moss, others use a handful of gypsum (land plaster). The land plaster, especially during warm weather tends to retain the ammonia in the droppings. The addition of either of these items help to keep the room free of odor and both improve the value to the droppings considerable.

Manure with which land plaster has been mixed forms a fairly well balanced fertilizer and can be used to advantage in gardening and farming. In various parts of the country large quantities of it are sold to florists and more than enough money is received to pay for the cost of removing it from the battery room.

The most modern method of cleaning droppings is by means of belts which replace the dropping pans entirely. This system is shown attached to the type of batteries described in the diagram and reduces the labor of cleaning to only a fraction of the time formerly required with dropping pans. These belts can be geared together so that all decks of the batteries can be cleaned at one time. The equivalent of from forty to sixty dropping pans can be replaced by one of these belt systems and if this is properly geared together the droppings from these belts can be removed in three or four minutes' time.

Considerable experimental work has been done in order to determine the proper type of belting material which will prove satisfactory for this work. Several different kinds of treated canvas have been used but the life of this material is very short owing to the fact that the ammonia from the droppings attacks it very quickly. The author has been working a great deal during the past six months in connection with one of the large rubber companies and has devised a combination belt of vulcanized rubber compound on heavy canvas. Considerable experimental work brought out the fact that pure rubber was readily attacked by several of the chemical combinations present in droppings of various ages. It was further necessary to incorporate several neutralizing oxides into the rubber combination in order to completely neutralize the effects of the ammonia and other chemicals present in the droppings. The resulting product obtained through this experimental work has been proving extremely satisfactory wherever it is being used for commercial work.

Poultry droppings are becoming better known each day for their value as fertilizer. One operator sold his year's accumulation for $900.00. This amount was sufficient to allow him to employ help in his battery plant to do all of the manual labor necessary. Another operator very recently informed me that he was delivering his daily supply of droppings direct to a well known florist and each month receives a check for $50.00. He advised me that this amount took care of all of his overhead expense such as rent, light, and heat, and allows him a much larger net profit on his broilers.

REMOVING DEAD CHICKS

In spite of all precautions some chicks will die. The sooner they are removed the better, especially if they are diseased. To a great extent, the death loss in battery brooders will occur in the nursery room. It will also stop there-with the proper type of chicks. While feeding or watering, it is well to scrutinize carefully the wire bottom on each tray of each battery. As soon as a chick dies the other chicks walk over it and soon press it into the wire mesh of the floor.

If not removed promptly each morning, these dead chicks soon rot and make a bad odor in the nursery room. To allow an accumulation of dead chicks, therefore is not sanitary nor the easiest job to clean up, as they are almost impossible to get out complete after they start to decay. On the other hand, if they are removed each day, they are easily pulled out with a wire hook.

It is of no advantage to put off this job from day to day. No disinfectant will destroy the odor. In many battery plants the operators are very careless about removing dead chicks. Some of them allow the chicks to remain three or four days on the battery floors, others let them lie on the floor or in the corners of the room. There is no excuse for such disgusting slovenliness.

If you keep accurate record of mortality, as you certainly should, it is easy to throw them into a small pail as you make your rounds, to remove them from the house, count them and bury or burn them.

Several healthy chicks may easily become weak or diseased or poisoned through contact with dead chicks. For this reason, if for no other, dead ones should be removed promptly and buried or burned.

Many operators have trained cats to live in the battery rooms to destroy the mice that get at the feed. Such cats can prove to be a big asset but the cats can be ruined for their work by allowing them to eat dead chicks. If you tempt even a well trained cat to eat dead chicks, it will not be long before it will decide to try live ones that escape from the battery trays.

While the death loss will be the greatest in the nursery room, there is still a possibility of older birds dying, even up to market age, so it is well when making rounds to scan each tray carefully for dead or crippled birds. In the later stages of brooding, it is especially important to remove a dead bird quickly.

In many cases cannibalism has started in the batteries because of the death of a bird near the outside edge of the compartment. The birds peck the dead one, get a taste of flesh, then pick on the weaker ones until these are killed—and eaten. It is possible for an entire tray of birds thus to acquire the cannibalistic appetite.

REMOVAL OF PULLETS TO RANGE

In various localities opinions differ as to the length of time advisable to keep pullets in batteries. Most operators keep the birds there only until the sex can be determined. This allows them to use their entire range for the production of pullets only. At present writing this seems the most satisfactory system to use.

White Leghorn pullets need be kept in the batteries for only two or three weeks, as at that time they can be recognized. If to be transferred then and if they have been handled under the temperature schedule outlined earlier in the chapter, they will be hardy and easy to manage. They will require heat at least until they become acquainted with their new quarters and especially during cold weather.

It is an advantage to have temporary roosts available and ready for them the first day they are removed from the batteries. Their previous training in the battery has taught them to be up off the ground, so they will take to the roosts quickly and easily. In some cases where the roosts are not available until the second, third or fourth day after the birds were brought into their new quarters, there has been difficulty in getting them on the roost. Battery bred pullets take easily to wire porches in front of their colony houses. If they are placed in houses with such or placed in range shelters, it is a good plan to chase them outside on the very first warm, sunshiny day so as to acquaint them with the fact that they may go out.

Some operators have found it satisfactory to wait until the pullets were six to ten weeks old before removing them from the battery brooders and then to take them directly to the laying houses. Thus they never come in contact with the ground. Contaminated ground cannot do them good and it may do them harm by spreading disease among them.

Some big egg producing birds were pullets kept twelve weeks in the batteries and then transferred directly to laying houses where they matured. Before attempting this procedure on a large scale try it out first thoroughly with a small number of birds.

Chapter 10. Other Brooders

10.1. Thermostatically Controlled Electric Brooders

Thermostatically controlled electric brooders use a heating element controlled by a thermostat. The heating element is attached to a hover made of sheet metal, usually with four built-in legs to allow it to stand on the floor, though the unit may be designed to be suspended from the ceiling.

Such brooders represent a technology which reached its full flowering in the 1930s and has since faded. You can no longer buy 250-chick insulated brooders with built-in curtains and tiny internal circulating fans. See Figure 1. The only currently manufactured thermostatic floor brooder I am aware of is an uninsulated 100-chick brooder from GQF (See Figure 2). However, there are probably tens of thousands of old brooders in barns and chicken houses all over the country.

Figure 1. An typical electric brooder from the Forties, designed for cold-house brooding: heavily insulated and with a curtain all the way around. Such brooders are no longer manufactured. (Lyon Rural Electric Co.)

The electrical circuit is very simple, consisting of a heating element, a thermostat (consisting of a thermostat wafer, a switch, and an adjusting bolt), and a power cord. Usually there are also one or two lamp sockets for attraction lights. If there are two, one is on all the time, and the other turns on and off with the heating

Figure 2. A 250-watt GQF floor brooder I bought several years ago. This has a high-quality heating element, but its uninsulated design is unsuitable for cold-weather brooding,

element. This is convenient when adjusting the brooder, because it gives you a visual indication of whether the heating element is on or off.

Thermostat wafers and switches sometimes go bad, but replacements are readily available from GQF and others. Replacement power cords can be obtained at any hardware store. Defunct heating elements can usually be repaired, since they consist of nothing but a length of nichrome wire (which provides the heat) held to insulators with screws. Usually, when the element goes bad, it's because one of the ends has come adrift from a screw. Either the wire breaks right at the screw or the screw rusts away. Nichrome wire is stainless, but the other metal parts usually aren't. It's usually possible to free the seized parts with Liquid Wrench or to abandon them and create a new connection point.

Warning!

Loose ends of nichrome wire will cause a shock hazard if they touch the sheet metal parts of the brooder. If you're testing one of these at a farm sale or something, think twice before plugging it in, and don't let anyone touch it while it's plugged in. You will want to inspect the heating element before placing the unit into service, to make sure that the wire is connected everywhere, and that it's all properly routed so it can't touch the sheet metal.

Any metal-shelled electrical appliance needs to be grounded. Brooders need a three-prong plug with the ground plug firmly connected to the sheet metal. All brooders made in the last thirty years will have this, but old ones will need to have their power cords replaced.

A three-prong plug doesn't provide any additional safety unless the circuit is actually grounded. Over the years, many homeowners have installed three-prong outlets in their homes without actually hooking up the ground connections. Someone did this to my house at some point. It's wise to spend a few dollars at the hardware store on a household circuit tester, which looks like a three-prong plug with three or four indicator lights on the end.

A ground-fault circuit interrupter (GFCI) outlet can provide shock protection even on a circuit that doesn't have a ground connection, and they are widely used in retrofitting circuits where there is no ground wire. However, GFCI circuits have some tendency to "false-trip," where they turn the circuit off even though there isn't a circuit fault. This is not something you want to happen to a brooder full of day-old chicks. On the whole, it is probably safer (for the chicks, not you) if you forego the GFCI protection, provided that the sheet metal of the brooder is connected to the ground wire of the three-wire power cord, and the circuit the power cord is plugged into has a good ground (as revealed by the circuit tester). Don't guess: test.

Any library will have several teaching novices how to perform these tasks, but if electrical wiring makes you uncomfortable, paying an electrician to do them will be money well spent.

Generally, the heating elements do not get hot enough to glow visibly, though they may have a dull red glow. This means that, unlike electric lamp brooders, they do not have a powerful light that attracts the chicks.

In any brooder that lacks an attraction light, place the brooder off-center in the draft guard, so the draft guard touches (or almost touches) the brooder at one point. Lost chicks will tend to follow the draft guard around in a circle, and will eventually reach the brooder.

10.2. Propane and Natural Gas Brooders

The mainstream poultry industry runs on propane or natural gas brooders, which are generally cheaper to run than comparable electric brooders. See Figure 3.

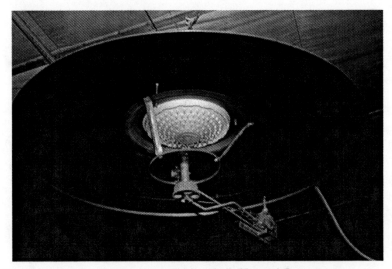

Figure 3. A modern propane "pancake brooder" (Univ. of Georgia).

The brooders themselves are simple, relatively inexpensive, and trouble-free. Many use mechanical thermostats that do not require electricity, which make them especially desirable in areas subject to long power outages. If you find yourself brooding 500 chicks at once and wish to use warm-house brooding, propane brooders are probably the way to go. Because today's propane brooders are uninsulated, they are not cost-effective for cold-house brooding. In warm-house brooding, the heat that escapes the uninsulated, curtainless canopy goes to warm the insulated brooder house. (Fifty years ago, you could buy insulated propane brooders for cold-house brooding. See Figure 4.)

Most modern brooders seem to come in the 30,000-40,000 BTU/hr. range, brood between 1,000 and 1,500 chicks, and, according to manufacturer's guidelines, should be used in houses providing about 800 square feet of floor space per brooder.

Figure 4. An insulated propane brooder from the Forties, with an unusual metal canopy with built-in pop-holes for the chicks. Such units are no longer manufactured. (Beacon Milling Co.)

30,000 BTU/hr. is equivalent to 11,720 watts, or about eight gallons of propane a day. Brooding 1,000 day-old chicks with insulated electric lamp brooders would require 1,250 watts in warm weather and 2,500 watts in freezing weather, so a modern propane burner running flat out uses 5-9 times the energy of an insulated lamp brooder. Many designs only modulate between half power and full power, so today's commercial propane brooders make sense only if you want to (and can afford to) heat the entire brooder house.

General-purpose **infrared** gas/propane heaters can also be used for brooding. **Many of these** are available in smaller sizes (down to 3,000 BTU/hr.) **and** give a more directional heat beam, which makes them **more efficient**. One manufacturer says that 1,000 chicks can be brooded with a 16,000 BTU/hr. heater, so presumably the same **manufacturer's** 3,000 BTU/hr. heater would handle about 180 chicks and use 0.8 gallons of propane per day running at full power.

10.3. Sunshine Brooders

Sunshine brooders are essentially an outdoor battery brooder. There is a weatherproof, insulated brooder chamber attached to a roofed wire run. The entire unit is generally put up on sawhorses for easy care. The brooder section may have either a litter floor or wire. These brooders, now forgotten, were very popular among egg farmers in the 1940s and 1950s, who often raised 10,000 or more chicks per year in them.

Figure 5. Old-time sunshine brooders (Lyon Rural Electric Co.)

While fresh air is the nominal reason for having them, I think farmers really liked them for the compelling reason that manure handling was less troublesome than with battery brooders. You simply let the manure pile up on the ground underneath and dealt with it when you got around to it. Sprinkling lime or superphosphate fertilizer on the manure will keep the smell down and the flies away. Some farmers built open sheds to shelter their brooder area from the weather.

10.4. Oil, Coal, and Wood Brooders

Brooders burning oil, coal, or wood were also used. I doubt that anyone looks back on them with nostalgia, because they required a lot of care compared to propane and electric brooders.

The original brooders used kerosene lamps for heat, and were good for about 50 chicks. The lamps had to be checked several times a day, and disassembled, cleaned, refilled, and trimmed once a day.

Coal brooders replaced kerosene in the 1920s. A single coal stove would brood up to 250 chicks, but would use over a thousand pounds of coal per brood.

Wood-burning brooders were much less popular than other types. Gasoline brooders had a brief burst of popularity around 1910, in spite of the obvious fire hazard. I hope we all know better today.

10.5. Fireless Brooders (Philo Boxes)

Several kinds of "fireless brooders" were used, which were basically heavily insulated boxes that conserved the chicks' body heat, so they could stay warm without an external heat source. The most common of these was the "Philo box." I have not investigated these closely, since they were considered by old-timers to be even more troublesome than kerosene-lamp brooders, which is saying something. The principle survives in the boxes used to ship baby chicks.

10.6. Hot Water Brooders

Many farms have used hot-water brooding, where the water was heated in a conventional central heating system and piped to the brooder houses. Lengths of copper pipe were used as radiators. Hot water heat is very effective, though it is important that the system be designed so the pipes can be lifted out of the way for house cleaning.

Mainstream poultry farms are looking into hot-water heat again, since combustible waste products (biomass) are very cheap, and reliable boilers are available that will burn wood chips, sawdust, or ground corncobs. In fact, poultry litter makes an excellent fuel if it isn't too wet. For people who can't find enough farms that want the manure as fertilizer, burning the litter at least keeps it from polluting streams and groundwater, in addition to saving money.

10.7. Floor Radiant Heat Brooders

Floor radiant heat (slab heat) is another technology that never quite goes away. Slab heat uses a concrete-floored brooder houses with embedded hot-water piping to heat up the floor, which provides all the heat necessary to brood the chicks. (I have also heard of recent efforts using piping buried in a dirt floor.) The heated area stretches from end to end of the house, but not all the way from front to back. Usually it occupies the central third of the house. Sometimes it is all the way at the back. In some designs, a long hover stretches the full length of the house, over the heated area. Slab heat has the disadvantage of heating and cooling very slowly, which is a problem in changeable weather.

Electric heating cable of the type used in greenhouses can also sometimes used to heat the floor. There was a period in England, when off-peak electric rates were introduced, where electrically heated floors were popular. The heating cables were covered with sand and the sand was covered with concrete paving stones. The concrete was heated at night, when electricity was cheap, and generally stayed warm all day without additional power.

With slab heat, sand is usually used instead of shavings as litter, since it conducts heat better than other litter material. Sometimes no litter at all is used.

One advantage of heated floors is that the houses are always very dry, which eliminates a lot of problems, such as coccidiosis. However, dryness leads to a lot of airborne dust, especially when sand is used as litter.

I have often wondered if a small amount of floor heat (using heating cables) would be a good idea in ordinary brooder houses, to keep the litter dry and to make the chicks more comfortable. In my climate, it is difficult to keep the brooder house as dry as it ought to be.

Chapter 11. Feeding

Feeding chicks is really very simple; you go down to the feed store and buy a sack of chick starter, then feed the chicks as much of it as they want. Chick starter provides baby chicks with a balanced diet.

It is possible to do without commercial chick starter, but commercial chick starter is cheaper, more convenient, and better than any readily available alternatives. I've always had good results from commercial feed. I've tried the feed from Purina, Nutrina, Land O' Lakes, Dairygold, and Kropf.

I've heard a lot of people go out of their way to badmouth commercial feed. I'm not going to argue about it. I'll just point out that there's a big difference between being able to find fault with someone else and being able to do a better job yourself. Formulating and mixing poultry rations is an advanced subject that you shouldn't tackle until after you've mastered the fundamentals of poultrykeeping.

For myself, I've read several books on poultry nutrition and feed formulation, and this has cured me of any desire to mix my own feeds. Feed science is pretty complicated.

11.1. Medicated or Non-Medicated Feed?

Chick starter can be divided into two categories: medicated and non-medicated. Medicated starter contains a coccidiostat; a drug to keep the chicks from coming down with coccidiosis. There are a variety of coccidiostats. Amprolium is the favorite with small flockowners, since it does not kill the coccidia outright, which means that the chicks can develop immunity while still being protected by the drug. After the chicks are switched from medicated chick starter to a non-medicated grower or layer feed, they are less likely to suddenly come down with coccidiosis.

If your management techniques make coccidiosis unlikely, you should use non-medicated feed (see Chapter 15). Otherwise, you should use medicated feed.

On our farm, I use medicated feed with my pullets, while Karen uses non-medicated feed for her broilers. The difference is that Karen raises broiler chicks on a rigid schedule that puts them out on range at 2-3 weeks. This is young enough that an outbreak in the brooder house is unlikely. My egg-type pullets, on the other hand, need heat longer, putting them into the danger zone, and I also have a track record for keeping them in the brooder house longer than I ought to. This puts them at risk for coccidiosis, so I use medicated feed. Once the pullets are out on range they get the same non-medicated poultry feed that the hens get.

Some people make fear-mongering claims that chicken feed is full of hormones, but this is not true. The only hormone ever used in any quantity was DES, which was banned in 1959. In any event, it wasn't put in feed, but was used by injection.

Antibiotics are often added to the feed used by the big commercial poultry flocks, but the only medication I've ever seen in chicken feed at a feed store is a coccidiostat.

The tag on the feed sack will always announce whether the feed is medicated or not, and, if so, what the medication is.

11.2. Chick Starter vs. Broiler Starter

Another division is between broiler feed and everything else. Broiler chicks grow a lot faster than other chicks, and their nutritional needs are enough different that they do better with specially formulated feeds. Often the broiler feeds don't have the word "starter" in their names, but have names like "Meat Builder" or "Meat Bird Complete" and are billed as the one feed you need from day-old to slaughter. A broiler starter should have a protein level of 20-22% or even more.

Traditional chick starter is used for everything else. It will probably have lower protein and lower energy than the broiler starter, but that's okay. A protein level of 18% to 20% is typical and appropriate.

11.3. Don't Let the Feeders Go Empty

Keep feed in the feeders at all time. Chicks need to eat many times per day, and they do better if there's feed in the feeders whenever they go looking for a snack. Trying to save money by witholding feed is, at best, a game for experts. There's an old farm saying: "You can't starve profit into a cow." It's true of chicks as well.

When the feeders run empty, the chicks will rummage through the litter looking for something to eat. The increased intake of feces that accompanies this may lead to an outbreak of coccidiosis. Hungry chicks also peck at each other more than they would otherwise, and you might have an outbreak of cannibalism on your hands.

There are two opposite approaches to keeping the feeders full. One is to use trough feeders and to refill them without fail twice a day. The knowledge that the chicks will go hungry if you break your routine helps you stick to a rigid schedule, at least in theory. The feeding routine tends to put you in closer contact with your chicks and the extra time in the brooder house make you more likely to notice problems as soon as they occur.

The other approach is to use high-capacity feeders such as tube feeders and fill them only as necessary, possibly only once or twice a week. This, along with automatic waterers, makes it possible to care for the chicks without adopting a rigid schedule and minimizes the time that must be spent in the brooder house. This has its good points and its bad points.

A rigid schedule probably allows a higher quality of care, but most people are not in a position to plan their lives around their chicks. However, spending time regularly in the brooder house is one of the keys to success.

11.4. Chick Scratch

Chick scratch is a mixture of cracked or ground grains that does not represent a complete diet, since grains are deficient in vitamins, minerals, and protein.

In general, baby chicks have a strong appetite for protein and will pretty much ignore chick scratch. However, as they get older their protein needs decline, and they will become more and more interested in grains. Usually, though, this process has hardly begun by the time the chicks leave the brooder house, so there's not much point to giving them scratch feed.

There are two important exceptions to this: the first 48 hours and as emergency backup feed.

11.4.1. The First 48 Hours

Chicks can suffer from a condition called "paste-up" during the first few days, where feces will adhere to the down on their rear ends and plug up their vents, leading to a peculiar form of constipation.

Opinions vary as to the seriousness of paste-up. Many hobbyist publications insist that it represents certain death, while poultry scientists insist that it causes little or no mortality. It's safe to say that the condition doesn't do the chicks a bit of good.

Paste-up is aggravated by chilling and by certain feed ingredients such as soybeans, which are a major component in chick starter.

Paste-up is the least of the problems caused by chilling, so fixing the cause takes top priority. But sometimes the chicks have suffered from chilling in shipment, which is out of your control.

By feeding chick scratch instead of chick starter for the first 48 hours, paste-up can be minimized. This has little or no nutritional downside, since the absorbed yolk is still providing the chicks with the vitamins and proteins that grains lack, and in any event the big challenge of the first 48 hours is to get water and calories into the chicks. Scratch grain has plenty of calories.

A compromise plan that has worked very well for me is to use chick scratch in the first feeders, but to install hanging tube feeders full of chick starter on Day 1. The first feeders should be placed closer to the brooders than the tube feeders, as usual. The chicks will discover the scratch feed first and the tube feeders later, which will give the effect you want.

If you're quite sure the chicks have been chilled, don't install the tube feeders until 48 hours after the chicks arrive. Otherwise, giving both kinds of feed in their different feeders will work fine.

Because they like chick starter better than scratch feed, they will be more enthusiastic about the tube feeders than they might otherwise be, and should transition from the first feeders sooner. This gives a second advantage to scratch feeding.

Continue feeding scratch feed in the first feeders for 5-7 days if the chicks are eating well from the larger feeders. That is, if you see lots more chicks eating from the larger feeders than the first feeders, continue with the scratch. If the chicks don't seem to prefer the larger feeders yet, switch to chick starter in the first feeders after 48 hours. Sometimes the chicks have to grow some more before they can manage the larger feeders comfortably. Don't try to force them.

11.4.2. Emergency Backup Feed

After the first week, you will probably be left with a sack of scratch feed that is still mostly full. It will go stale by the time your next batch of chicks arrives, so it's probably best to install an extra feeder in the brooder house and put the scratch feed in it.

The chicks will largely ignore this feed unless the other feeders run empty, or if there's something wrong with the chick starter. The latter case is very rare (it has never happened to me), but it happens once in a while.

In the normal course of events, the amount of scratch feed will decline almost imperceptibly day by day. But if the other feeders go empty, the chicks will eat scratch feed instead. By providing an emergency backup feed, you will reduce the amount of litter-eating and feather-picking that might otherwise occupy the time of hungry chicks, and this will in turn reduce the chances of the empty feeders triggering an outbreak of coccidiosis or cannibalism.

With luck, all the scratch feed will have been eaten by the time you move the chicks out of the brooder house. If not, move it with them.

11.5. Tube Feeders

Figure 1. A small hanging tube feeder suitable for chicks. Similar units are available at any feed store.

There are two basic kind of feeders: tube feeders and trough feeders. I prefer tube feeders.

Tube feeders come in two sizes: large and small. The small ones hold about fifteen pounds of feed and are the type you want for the brooder house. The large ones have a much deeper feed pan and are suitable for adult birds, but are not appropriate for chicks.

The height of the tube feeder needs to be adjusted every few days as the chicks grow. If the feeder is too low, litter gets scratched into it and the chicks waste a lot of feed. If the feeder is too high, the chicks can't get enough to eat.

Adjusting the feeder is very easy if you use a simple wooden toggle as shown in Figure 2. You can make a lifetime supply of such toggles in a few minutes if you have a drill, a handsaw, and a length of 1x2 in. lumber.

Figure 2. A simple toggle makes height adjustment easy. This is simply a 6 in. length of 1x2 lumber with two holes slightly larger than the rope. The end of the rope is knotted to keep it from pulling through. Grab the toggle, lift the feeder to the desired height, and release.

11.6. Trough Feeders

There are a lot of good chick troughs on the market. You have your choice of plastic or galvanized steel. If you use troughs, you need two sets in the brooder house: one for young chicks and one for older chicks.

The starter troughs made for young chicks will be outgrown surprisingly quickly. My experience is that broiler chicks can no longer stick their heads through the holes in top-hole slide-lid troughs by the time they are two weeks old. Before this happens, you should have already installed larger troughs to give the chicks time to get used to them, so they will not have to undergo a major readjustment when the small ones are removed.

Figure 3. An old-time ad for galvanized slide-top starter troughs and quart-jar waterers. Identical metal products are still made today, along with similar plastic ones.

Starter troughs should be set on the ground, but it's best if larger troughs are hung from the ceiling. This make it easier to adjust the height and also keeps mice and rats out of the troughs once the height has been adjusted beyond their reach.

11.7. Grit

Probably everyone knows that chickens store rocks in their gizzards to grind their feed, as a kind of artificial teeth. The value of grit in the brooder house is questionable, since the chicks are generally being fed only chicken feed and perhaps cracked grains, which require little additional grinding. Even when fed whole grains, grit is not absolutely necessary.

The chicks like grit and will be delighted to eat it if you provide it. You can buy chick grit at any feed store. You can feed as much as you like to the chicks. They won't hurt themselves by over-indulging. Old-time poultrymen kept a separate feeder full of grit topped up at all times. Probably the most common feeding method these days is to sprinkle a handful of grit on top of the regular chicken feed once a week or so.

I used to buy grit sometimes, but I don't bother anymore.

Grit is more important to older birds who are being fed whole grains.

11.8. Supplemental Feeds

I don't feed table scraps or other supplemental feeds to chicks in the brooder house. This is another task that I think is best deferred until the chickens are larger. My hens do a much better job of disposing of such things than chicks, and my brooder houses are already so cluttered with brooders, feeder, and other equipment that finding room for supplemental feeding would be something of a trick. The chick also eat relatively little, so there's more of a chance that scraps might be left uneaten and have time to go bad.

If you do feed table scraps or other occasional feeds, feed them in a trough. Don't toss them on the floor. The next time you visit the brooder house, remove any uneaten portion and, if the feed was wet, take out the trough and rinse out any bits that might be adhering to it. Chicks aren't as rugged as adult chickens and we don't want to take any chances on potentially spoiled feed.

11.9. Range

Chickens find lush green range palatable, and range can provide growing chickens with all of their vitamin needs and much of their protein needs, though their diet will still be deficient in calories and minerals unless you raise a very small number of chickens in a very large space. However, lush green range is seasonal, and is at its most nutritious in the early spring when it is too cold and wet for little chicks to venture outside.

Back before anybody knew what vitamins were, it was known that chickens needed green feed and sunshine if they are going to thrive. Green feed provides every vitamin but vitamin D, which chickens, like people, can synthesize with the aid of ultraviolet light on their skin. Thus, good poultrykeepers went to a lot of trouble to get the chicks outdoors on lush green range as early as possible, urging them out at two or three days of age if the weather was perfect, and shooing them back inside as soon as the weather looked threatening.

These days, chicks can grow perfectly well in confinement using readily available commercial rations, and one does not have to keep an eye on the sky throughout the brooding season.

Not that I have anything against range. I love range. On our farm, all the chickens go on range as soon as they leave the brooder house—but not before then.

It's not just the weather. Chicks are more at risk from predators if they have access to the outdoors. They can be attacked when they're outdoors, and the predators can come in and make themselves at home if there's an open door.

Another reason is that chickens are very hard on range. They scratch the turf to pieces looking for bugs, worms, and seeds, and any yard attached to a permanently sited chicken house soon becomes a sea of mud (or dust, depending on the season). Such a yard does not promote healthy chicks.

To me, the only sensible approach to range is to use portable houses, which get towed to a new location when the old one starts getting muddy. In the old days, this is exactly what farmers did with their brooder houses. I find it far more convenient to have the brooder houses close to my home, where I can keep an eye (and an ear) on things, and brood the chicks in confinement. They'll be on range soon enough.

Chapter 12. Waterers

The labor of watering poultry by carrying water in buckets is tremendous and not to be considered on any up-to-date poultry plant.
—Milo Hastings, *The Dollar Hen*, 1909

One of the worst things that can happen to your chicks is for them to run out of water. Equally bad is for their brooder house to become soaked with spilled or leaking waterers.

Though I start each batch of chicks with quart-jar waterers, I switch to automatic watering systems as soon as possible. Automatic watering systems are not completely trouble-free by any means, but I find they deliver water more reliably than I do.

In my experience, if you fill the waterers by hand, watering the chicks will be your most time-consuming chore. With automatic waterers, there will be periods of essentially zero labor punctuated by frantic repair efforts when some part of the system fails. In either case, water will be one of your biggest concerns and needs to be taken very seriously.

Even if you fill your waterers by hand, you will want water piped to the brooder house, if only in a garden hose left outside the door. Better would be a laundry tub. Since broilers do not fly, you could put such a tub inside the brooder house and save yourself some steps. Other breeds fly all too well and an inside sink would be a bad idea. A laundry tub on the outside of the brooder house would still be a great convenience.

12.1. Metal-Valve vs. Plastic-Valve Waterers

I like to divide automatic waterers into two categories: metal-valve waterers and plastic-valve waterers. Metal-valve waterers can withstand ordinary household water pressures and don't crack when they freeze. Plastic-valve waterers require a pressure reducer before they can be operated from household water lines, and they will break if the water inside them freezes. Some of the most convenient and interesting waterers are plastic-valve waterers. I use metal-valve waterers because I don't like having my brooder-house flood after a cold night.

12.2. Prevent Wet Litter

Whichever kind of waterer you use, you need to protect against wet litter around the watering area. If you put a screened stand of some kind under the waterer (which you should), you can put a pan under the stand to catch any spilled water. Sticking a hose in the side of the pan and running it outside will help keep the pan from overflowing, but everything in a chicken house clogs with litter eventually, so this is more of a convenience feature than an anti-flood device.

I've seen photos of houses that old-timers had modified to have a special wire-floored drinking area that sticks out beyond the house, so any spilled water falls outside the house. This is supposed to eliminate a lot of problems. You'd want some kind of draft-proofing on it when the chicks are little.

Preventing wet litter also means that the water level inside the waterer shouldn't be too high, and chicks must be prevented from roosting on the waterers. Roosting will break or knock over the smaller waterers and will cause larger ones to tip and spill.

12.3. Keep it Clean

Many experts tell you to clean and sanitize your waterers every day. I have trouble believing that anyone except a few hobbyists can find the time to do this. Personally, I rarely sanitize any of my brooder-house equipment. It hardly seems necessary because I rarely see a sick chick and never see a sick flock.

What I do is clean the waterers between flocks, and clean them whenever they look dirty. If rinsing makes them look clean, I just sluice some water through them. Otherwise I wipe them with a paper towel.

The main thing is not whether you use chemicals or not, but looking at the waterers every time you go into the brooder house, and cleaning any waterer that looks dirty. You don't want crud to sit around in the water for any longer than necessary.

When chickens eat, the feed goes into their crops first. The crop is just a pouch in between the throat and the stomach, and it doesn't have a valve on the top. If the chicks have to bend down to drink, some of the stuff in their crops will come out into the water. Clean water demands that the water level be higher than crop level to keep this from happening.

If the chicks can roost on or above the waterers, the waterers will get a lot of manure in them. To prevent this, you need some kind of roosting barrier or a different kind of waterer. In particular, you can't string water hoses through the air in a chicken house (unless you have non-flying broilers), because chicks will perch on the hoses. The hoses need to be where the chickens can't perch on them—tied off to rafters, for example.

12.4. Manually Filled Waterers

Most manually filled waterers are vacuum founts, just like the quart-jar waterers, only bigger. There are one or more holes in the reservoir, and when the water level in the bowl drops far enough, the holes are uncovered, air bubbles up into the reservoir, and water pours down. When enough water comes down that the holes are covered again, no air can bubble up, so no water can come down.

The smaller units have a screw-on base. You turn them upside down to refill them. The larger ones are double-walled galvanized units with an inner and an outer shell. You take off the outer shell and fill up the inner one. These are somewhat freeze-resistant due to the double-wall construction.

Usually, feed stores carry plastic one-gallon waterers and galvanized waterers in the larger sizes (2-5 gallons), but these larger sizes are available in plastic as well. I've never used plastic waterers in sizes larger than one gallon. I've used the larger galvanized waterers. I don't like them. The water always seems to be fouled with manure and they start to rust after only a year or two.

These waterers should be put up on some kind of stand to keep litter out of them. This would typically be made with a rectangle of 2x4's covered with ½ in. hardware cloth or welded wire. You can't hang these waterers from the ceiling successfully, which is one of the reasons why I don't like them. To adjust their height, you'll need an assortment of boards or bricks to put under them, increasing their height above the stand.

12.5. Little Giant Waterers

The most ubiquitous automatic metal-valve waterer is the Little Giant bowl waterer sold by Miller Manufacturing, which is probably available in every feed store in the country. These waterers have been in continuous production for over fifty years.

Figure 1. One of my Little Giant waterers.

12.5.1. Advantages and Disadvantages

Their advantages are that they never seem to be damaged by freezing, they are readily available, they work at any water pressure, and they last forever if you don't step on the plastic water bowl. Their disadvantages are that they are difficult to adjust and tend to disassemble themselves under the strain of being pecked at continuously by thirsty chicks if the water ever fails.

12.5.2. How it Works

The waterer's valve is opened by a spring when the bowl is light. As water fills the bowl, it becomes heavier and overcomes the pull of the spring, and the valve is closed. The valve is adjusted by a nut that tensions the spring. A second nut is used to jam the first one. In theory, this prevents the first nut from unscrewing itself.

12.5.3. Failure Modes

- If you don't use two pairs of pliers, one for each nut, and jam them together good and hard, the waterer will go out of adjustment.
- The nuts and spring are covered by a brass stem that attaches to the bowl. If the bowl unscrews itself from the stem, your brooder house will flood. (Ask me how I know.) Make sure the stem is screwed into the bowl good and tight.
- If the stem falls off the upper part of the waterer, it (and the bowl) will fall to the floor. This leaves the valve closed, so the house won't flood, but the thirsty chicks will peck at the exposed spring and nuts, and if you didn't use two pairs of pliers to tighten them like I told you to, the nuts will quickly become unscrewed. Both nuts and the spring will be lost in the litter.
- There is a screen inside the waterer to prevent larger particles from getting into the valve, which can jam it open. This screen will clog eventually.

12.5.4. Preventative Maintenance

In spite of this impressive list of things that go wrong, I use these waterers all the time. Really, I don't have a lot of choice, since most other automatic waterers self-destruct if they have water in them when there's a good hard freeze. The low-cost plastic waterers were developed for the mainstream commercial poultry industry, which uses heavily insulated houses and warm-house brooding. By the time the chicks are large enough that they don't need heat, their body heat alone is enough to keep the house above freezing. I don't have an insulated, controlled-ventilation brooder house, so I'll keep using metal-valve waterers.

The following tips will make it so that most of the failure modes listed above are things that happen to other people:

* Before each batch of chicks, trigger the waterer manually to make sure that a good, strong flow of water is coming out. If you get an unimpressive dribble, take the unit apart and clean the screen.
* If your water isn't sediment-free, use a screen or filter to catch most of the particles before they reach your waterers. This will reduce the likelihood of the screen clogging shut or the valve sticking open.
* Always use two pairs of pliers when adjusting the waterer.
* Adjust the waterer to be only about half full. This gives you a margin of error and prevents rowdy chicks from causing a lot of spillage.
* It's easy to put the bowl/stem assembly on incorrectly. There's very little "feel" to it, and you can leave it in the wrong position very easily. Wiggle it around a couple of times until you're sure it's on right.
* Keep a spare unit in case of trouble, so you can get a non-working unit swapped in no time, and the birds don't have to go without water while you scratch your head over how to fix the faulty one.

12.5.5. Installation

The top of the waterers ends in a ½ in. female pipe fitting with a rubber washer and the brass screen. To make a low-pressure waterer, I like to attach the waterer to the bottom of a one-foot length of half-inch galvanized pipe. At the top I put a ¼ in. hose barb, which I attach to ¼ in. hose with a pipe clamp. I use a larger pipe clamp around the top of the galvanized pipe to attach a loop of wire or baling twine. I then hang the waterer from the loop on a length of rope or chain. You could hang the waterer directly from the hose if you wanted to. To keep it from kinking, you could hang it from something rounded, like a pulley.

For a waterer run off household water pressure, you can put a female garden hose adaptor at the top of the pipe. Garden hose is too stiff and heavy for the assembly to hang straight unless the section of galvanized pipe is about two feet long. Three feet is better.

Miller has recently started offering an installation kit with a wall bracket, which should simplify things.

The waterer's brass screen will become clogged if your water has sediment in it. It's more convenient to add an external Y-filter (sold in hardware and garden stores for use with drip irrigation systems) than to remove the waterer from the pipe to get at the screen. However, the Y-adaptors I've found are all plastic, so they'll freeze and break in cold weather. At the very least, put the Y-filter outside so the leakage from a broken one doesn't flood the house. A filter in you pump house is better.

Use one Little Giant waterer for every 50 chicks.

12.6. Bell Waterers

Bell waterers are large plastic waterers that work on the same principle as the Little Giant waterer, but have a plastic valve assembly that splits in a good hard freeze. They feature a bowl over a foot in diameter, meaning that a lot more chickens can drink from it at once. Bell waterers are low-pressure waterers that will fail at household water pressures. You need a pressure regulator. They are easy to install, since they typically come with a rope to attach to the ceiling and a length of ¼ in. tubing to hook up to your water system.

The water level in a bell waterer is pretty much non-adjustable, and for this reason they lack all the adjustment-related failure modes of the Little Giant waterers. Like the Little Giant, they have a screen that clogs eventually, though it generally can be removed without tools.

Use one bell waterer for every 65 chicks.

Figure 2. Bell waterers (Univ. of Georgia).

12.7. Cup Waterers

Cup waterers are very small—only an inch or two in diameter. They either have a tiny float valve or a trigger valve that lets a little water into the cup when a chick pecks at it.

I've used cup valves on my battery brooders, and they were very effective. The ones I used were trigger-style units from GQF, which have a 1/8 in. pipe fitting on the back. GQF also sells two kinds of adapters. One adapts the waterers to a hose barb and had a built-in clip to attach the waterer to cage wire. This one can be used with laying cages or battery brooders. The other adapter allows you to attach cup waterers to lengths of ½ in. PVC pipe. For example, you could put ten waterers on a five-foot length of PVC pipe.

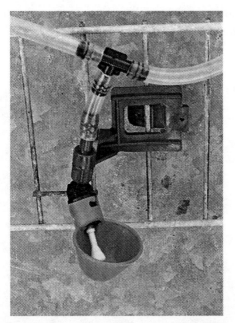

Figure 3. One of my cup waterers, of a type sold by GQF and others.

I have made only limited use of cup waterers. My biggest problem was leaks at the hose barbs, though this may have been due largely to my use of hardware-store vinyl tubing instead of the stuff sold by GQF especially for low-pressure water systems. In any event, if you use low-pressure waterers and plastic tubing, you should invest in a handful of tiny hose clamps to fix any dripping connections.

Because cup waterers are so small, only one or two birds can drink from one at a time. You need about one cup waterer for every ten chicks.

12.8. Nipple Waterers

Nipple waterers are all the rage at the big commercial installations. They are pretty much just a cup waterer without the cup. The waterer is placed above the chicks. When a chick pecks at the trigger, drops of water course down it and the chick drinks them.

Nipple waterers are simple, inexpensive, and self-cleaning. Like every other waterer on the planet, they have a reputation for getting stuck open and dripping, but they tend to drip slowly enough that farmers procrastinate replacing them, which leads to trouble.

Figure 4. A nipple waterer (Univ. of Georgia).

The thing I don't like about nipple waterers is that they need to be adjusted to exactly the right height if they are to work effectively. Also, the water pressure needs to be adjusted more carefully than with other kinds of waterers. Given the choice between cup and nipple waterers, I'd take cups.

12.9. Pressure Regulators

There are a variety of pressure regulators on the market. The cheapest is from GQF Manufacturing, which is basically a tiny propane pressure regulator with a garden hose connector on one end and a hose barb on the other. I used three of these last year. One failed after a season in sediment-laden water, and the others worked fine. These regulators worked very well with cup, bell, and Little Giant waterers. Keep a spare regulator in case one fails.

For nipple waterers, you'd want a regulator designed especially for the job, which costs more.

Chapter 13. Litter

Chicks can be brooded on wire or litter floors. Wire floors are used in sunshine brooders and battery brooders, while litter is used with conventional brooding.

I always use softwood shavings for litter. I usually buy them in compressed bales, though it's cheaper if it is bought loose and in bulk (at least a cubic yard). Hardwood shavings are supposed to be too dry and often too full of tannic acid for safety. Softwood sawdust, ground corncobs, peat moss, shredded sugar cane, peanut hulls, chopped straw, wood pellets, and even sand have all been used successfully for brooder-house litter.

Ordinary straw that hasn't been run through a chopper is generally considered to be too coarse for chick litter, though some people swear by it. Straw becomes matted easily in a chicken house, glued together by chicken manure. With mature hens, this can be avoided by feeding some grain in the straw, which causes them to spend hours every day scratching through it and fluffing it up. Chicks are too small and too uninterested in grain for this to work.

Start with at least four inches of litter in the brooder house. Start the brooder at least 24 hours before the chicks are scheduled to arrive to make sure the litter is warm and dry. If the brooder sinks into the littler, put bricks or wooden blocks under the legs, or suspend the brooder from the ceiling.

Every day, when you take care of the chicks, look for wet spots in the litter, especially near waterers, windows, and doors, and remove them. Also, keep an eye open for litter that is caked over with manure, and remove that. Caked litter is only a fraction of an inch thick, and you can lift it up in chunks the size and thickness of a dinner plate on a shovel or spading fork. In small brooder houses, a five-gallon bucket will hold a couple of days' worth of litter removal. Add fresh litter to the affected areas, or use a rake or hoe to even things out.

An alternative to removing wet or caked litter is to heap it up in a corner of the brooder house. The heap of litter will begin to compost, and the flakes of caked manure will fall apart and much of the water will evaporate. The inside of the heap will become quite hot, destroying many pathogens. The heap can be spread with a rake after it has become dry and friable again in a few days.

If caked litter becomes a big chore, it probably means that the the brooder house is overcrowded.

If wet litter becomes a big chore, the ventilation in the house is inadequate, there is a leakage or spillage problem with the waterers, or you are suffering from condensation on the walls and ceiling (which is cured in the short term with more ventilation and in the long term with insulation). Of course, if the brooder house has a dirt floor that is subject to flooding, wetness will be a problem whenever it rains. You should move or raise the brooder house in that case.

Wet litter is unhealthy for the chicks, since it chills them and promotes disease (particularly coccidiosis and roundworms). Wetness also rots walls and wooden floors in short order. The combination of water and manure is very destructive to wood and metal.

If the litter becomes very wet, it may be necessary to replace it. Some people expect to replace their litter more than once per brooding cycle, as the only way of controlling moisture in their brooder house. This is a backbreaking way of going about things. It is much better to control the sources of excess moisture (crowding, condensation, inadequate ventilation, leaky roofs, leaky floors, leaky waterers) than to deal with the consequences.

13.1. Built-Up Litter

During the farm labor shortages of World War II, a few researchers realized that a lot of farmers weren't changing their litter anymore, and that nothing bad was happening at all. In fact, chicks brooded on old litter rarely came down with coccidiosis, unlike chicks brooded on new litter. In addition, chickens of all ages on old litter did better on diets that had no animal protein in them.

This was the beginning of the heyday of built-up litter, also called "deep litter" and "compost litter."

The dietary effect of built-up litter is caused by vitamin B-12 that builds up in the litter as a side effect of bacterial action. Meat products have plenty of B-12, but plant protein products are deficient in it. As soon as B-12 was identified around 1950, feed mills started adding it to their poultry feeds and the dietary benefits of built-up litter became irrelevant.

13.1.1. Anti-Coccidial Properties

The anti-coccidial properties are more interesting, at least to those of us who prefer using non-medicated feeds. After litter has been used by chickens for at least six months, it acquires significant anti-coccidial properties. Presumably this is because microorganisms that eat coccidia have had time to take up residence in the litter and have multiplied to the point where coccidia have a hard time surviving.

The anti-coccidial effect of built-up litter is real but isn't entirely consistent. Some people have better luck with it than others. In any event, controlling wetness in the litter and keeping litter and manure out of the feeders and waterers is important for good results.

13.1.2. Labor-Saving Properties

Even if you aren't depending on the anti-coccidial properties of the litter, there's a lot to be said for not having to remove all of the litter between batches, which is a tedious chore. In my brooder houses, I remove no litter at all unless the old litter is very wet or is getting too deep—in which case I remove up to half the litter, but no more than that. This cuts the amount of work in half and retains whatever beneficial properties the litter might possess.

Used litter makes excellent fertilizer. I usually dump mine around my fruit trees in the back yard. I don't agree with the current fad for composting everything before use. It works just fine as is. If it's smelly or attracts flies, sprinkle some hydrated lime over it, and the problem will disappear.

13.1.3. Hydrated Lime

Speaking of hydrated lime, adding ten pounds of hydrated lime for every 100 square feet of brooder house (about a coffee can full) will make the litter looser and seem drier (though the moisture content isn't really changed). Also, coccidia don't like alkaline environments. After sprinkling it over the litter, stir it in with a rake or hoe. It's easier to do this when there aren't any chicks in the brooder house.

13.1.4. When to Give Up on Built-Up Litter

I would remove all the litter and disinfect the entire brooder house if I had an outbreak of some kind of scary disease that remained infectious after a long period in an empty house. So far, I've never suffered from such a problem. If I did, I'd ask for detailed instructions from the Extension Poultry Specialist for my area.

13.2. Flooding

Something that happens to me from time to time is flooding in the brooder house, usually due to a failure in an automatic waterer, though a leaky roof can be just as bad.

If you have actual standing water in the house, you need to bail it out. Once you're down to horrible soppy litter, you have the choice of shoveling it out or putting down new litter on top.

Putting down new litter on top of a soggy mess is more effective than you'd think. The last time I tried it, I was amazed at the difference one bale of shavings made. So don't despair if this happens to you. Shovel out as much litter as is convenient, and then put down plenty of new shavings. The new shavings are likely to stay acceptably dry on top.

Litter that has been soaked tends to become a compact, soggy mass that doesn't dry well on its own. Piling new shavings on top hides the problem, but you will eventually want to fix it. You can aerate it again by shoveling it into a heap. The newly aerated heap will drain, begin composting, and will eventually dry itself out. This is hard to do with chicks in the house.

Chapter 14. Brooder Houses

Like most people, I suppose, I have never actually built a brooder house. I have brooded chicks in modified outbuildings and in modified henhouses, but never in a house that was built from scratch specifically for brooding. But I'll cover the topic as best I can.

14.1. Floor Space Requirements

The traditional chart of floor-space requirements is as follows:

Age, Weeks	Square Ft. per Chick
0-4	½
4-8	1
8-12	2
12+	3

Figure 1. Floor space requirements

For broilers, I would go to 1 square foot per chick at three weeks, tops. The big commercial growers don't give their birds this much space, but more space is necessary if we want to avoid having the litter cake over. And we want this very much if we want to enjoy raising poultry. Caked litter is disgusting.

Chicks of all breeds, but especially broilers, grow remarkably quickly. Signs that they are outgrowing their quarters are, in order of appearance, caked litter, wet litter, feather-picking, and cannibalism. If you're using non-medicated feed, coccidiosis will probably appear somewhere in the sequence as well. If you give the chicks space when they need it, you can avoid a lot of trouble.

If you have unusually poor ventilation, you will need to give the chicks more room. There was a fad around 1900 for tightly sealed chicken houses. The lack of ventilation increased wetness and disease, and people using such houses found that they needed to give 8-12 square feet per hen for good results. With twentieth-century methods used from roughly 1910-1960, this falls to 3-4 square feet. With modern housing (with computer-controlled ventilation), it's around 1.75 square feet per hen. These numbers are for hens, but the principles are the same for chicks. A lot of this has to do with ventilation, but light levels and breed selection play a part, too.

The important rule to remember is that if you had poor results last time, try reducing the number of chicks next time. The odds are pretty good that your problems will vanish. With less crowding, the litter is drier, the birds are less stressed, and there's less of a crush around the feeders and waterers, not to mention under the brooder.

After a couple of years of disappointing results with 100 chicks per brooder house, I reduced my orders to 75. The difference was like night and day. Everything was easier; everything worked better.

You can sometimes get the same effect by moving the chicks out of the brooder house sooner. If they're crowded in the brooder house, they may be better off in their adult quarters even if, in theory, they ought to be getting heat for another week or two. For older chicks, giving up heat can be the lesser of two evils.

14.2. Brooding In a House for Older Birds

Probably the most common place to brood chicks is in the house they will occupy after the brooding period: a broiler house or a henhouse, depending on the type of bird. If we are talking about flocks of under 400 birds or so, the house is probably fairly small, with windows only on the front side, and a wood, dirt, or concrete floor with some kind of litter on top.

Making such a house ready for the chicks is largely a matter of removing unnecessary equipment such as nest boxes and adult-sized feeders, closing the windows and other openings, and installing brooders and draft guards.

Any leaks in the roof, broken hinges, missing window panes, and so on should be fixed now. You can get away with a lot more when dealing with adult birds than with day-old chicks, so you should always take stock before installing a new flock. There may be problems you've gotten used to and don't even notice anymore unless you make a special effort.

The old litter should be de-caked and any wet spots removed.

All the feed used with the previous flock should be removed. If you suspect rodents in the house, you should trap or poison them now. Rats will kill every one of your chicks if given the chance.

A layer of new shavings should be put down inside the draft guards. While the benefits of used litter are real, the chicks may eat quite a bit of litter before they fully understand the distinction between litter and food, and we don't want them to get a huge dose of whatever microorganisms are in the litter on their very first day.

If the house has windows only on the front, it's probably best to brood the chicks all the way at the rear of the house, because there will be less draft there. If it has windows front and rear, brood the chicks in the middle of the house.

Brooding chicks in the same house they'll live in as adults means that you will have plenty of space. Speaking as someone whose brooder houses are only eight feet square, this is quite a luxury.

14.3. Dedicated Brooder Houses

The ideal brooder house:
- Is high enough above grade to be floodproof.
- Is ratproof. It has a concrete floor and maybe even concrete walls, at least for the first foot or two. Second-best is a house with a wooden floor that's raised up on concrete piers about a foot off the ground, which also discourages rats.
- Is convenient to your home. You want to be able to check on the chicks several times a day without inconvenienc.e An

inconveniently placed brooder house isn't going to get as much attention as a convenient one.

- Is insulated. Insulation reduces the daily temperature swing and eliminates condensation on the ceiling and walls of the brooder house. It will reduce the cost of brooding and, at the same level of ventilation, provide a drier house.
- Is roomy. A cramped brooder house makes proper management difficult, and is an invitation to disaster if the chicks aren't moved out on time.
- Has proper plumbing and wiring. An ideal brooder house has a water supply that won't freeze and has wiring that's properly strung, fused, and grounded. The house should have area lights to make it possible to work inside at night, and it should have outlets for electric brooders, preferably weatherproof outlets because of all the dust you get in a brooder house.
- Has highly adjustable ventilation.
- Has self-closing doors.
- Has locks on the doors.

Most of us have brooder houses that are far from ideal, but we should do the best we can and try to make improvements whenever possible. My long-term goal is to replace my two tiny brooder houses with a single, larger house made from insulated concrete blocks. Such a house would be ratproof, rotproof, easy to clean, and would last forever.

14.4. Floors

14.4.1. Concrete Floors

The ideal floor is made of concrete. Concrete is rotproof, ratproof, and easy to clean—but expensive.

My brooder houses have concrete floors, but only because there was already a concrete slab on my farm in a convenient location.

14.4.2. Dirt Floors

A dirt floor works quite well if it's above grade and you can keep the rats out. I used dirt-floored brooder houses for years without any trouble. (My rat problems didn't start until later.)

In one dirt-floored brooder house, I put down wire mesh to discourage predators, then put in several inches of gravel to make sure the house wouldn't flood. In retrospect, I think the wire mesh was a waste of time, since it probably rusted away to nothing in less than a year.

14.4.3. Wood Floors

A wood floor is okay if it's kept about a foot off the ground. Any lower and the rats will feel safe living underneath the brooder house, and will quickly chew through the floor and into the house. With a foot of space between the ground and the floor, the rats don't feel safe, and cats and other predators have more room in which to work.
Wood floors set directly on the ground rot very quickly and should not be considered.

A wood floor is hotter in the summer and colder in the winter than concrete or dirt floors. Insulating the floor (for example, with a layer of foam board between two layers of plywood) will reduce this effect.

14.5. Walls

The main issues with walls are rot and rats.

14.5.1. Rot

Wet litter will ruin both wooden and metal walls. Single-boarded houses, with outside sheathing but no inside paneling, will rot quicker than double-boarded houses. To protect the structure, then, some kind of double-boarding is a good idea. I have never used double-boarding, but I'm paying the penalty; my walls are rotting out at the bottom.

The inner paneling can be exterior-grade plywood or some kind of plastic. Aluminized bubble insulation might be entirely adequate as a paneling material. I have never insulated a chicken house with it, but I've used it around chickens and know that they don't peck it to pieces, which they do with styrofoam. Aluminized bubble insulation is installed with a staple gun. It's a great vapor barrier and a fair to middling insulator. You can also buy a variety of vinyl-clad panels that ought to resist rot for a long time.

14.5.2. Rats

Rats have little difficulty chewing their way through wood or plywood walls, and can tunnel into a dirt-floored house. If rats can find a safe place from which to gnaw their way into a house, they will.

Eliminating places they find safe will help protect your chicks, even if you have a wooden-walled, dirt-floored house. Remove all equipment, trash, and bushes from around the house, and don't let the weeds grow tall. If the rats have to be in an open space to gnaw through the wall, they're far less likely to try.

Rats won't gnaw through concrete or galvanized steel. When I had a rat infestation last year, I excluded them from my brooder houses with strips of galvanized flashing nailed to the insides of the walls. I put these on all the sides where they had made a real effort to get in. Some sides were more exposed than others, and they never even tried to get through on the most exposed sides.

14.5.3. Wall Insulation

Besides aluminized bubble insulation, obvious possibilities are fiberglass and styrofoam. I don't like working with fiberglass, so I have had little experience with it. You can buy styrofoam insulation in 4x8' sheets and cut them to fit the gaps between the studs, or you can panel the inside of the chicken house with whole sheets. Chickens love styrofoam and will eat it if they get a chance, so you'll have to cover the foam up with something else. This protective cover only has to extend about two feet above the floor.

Exterior walls of wood or plywood should be painted. If the walls are to be painted, they should be painted when new. Once the wood gets weathered, it takes a lot more paint to cover it.

Galvanized steel makes an excellent exterior wall if it is protected from manure. Steel roofs seem to last at least thirty years, but wall should last even longer.

14.6. Roofs

I like galvanized steel roofing, but I don't like cutting it. Most lumber stores stock steel roofing in a few standard lengths, so you can avoid cutting if you design your house to use these lengths. Your supplier can also order roofing in any length you like, and the price is the same (at least, that's my experience).

In my houses, there is nothing under the metal roofing. No sheathing, no insulation. I don't even have rafters, just purlins. The roofing is nailed to the purlins with rubber-washer roofing nails, and that's it.

A more suitable roof for brooding would be insulated. An insulated roof not only leads to a warmer house, it leads to a drier house. This is because a cold metal roof will condense any moisture in the warm air coming out of the brooder, and it can drip back into the house. With an insulated roof, there's no condensation, and, assuming that there's some ventilation, when the warm air escapes, it takes all its moisture with it. And, of course, an insulated roof will lead to a cooler house on hot days.

Chickens don't like large daily variations in temperature, so insulation is a good thing. It's really just a question of how much you want to pay for it in time and expense. An insulated, double-boarded chicken house will take more than twice as long to build as an uninsulated, single-boarded house. The cost is likely to be more than twice as much as well.

14.7. Doors

Doors should start at least eight inches off the floor, or deep litter will interfere with their operation. Otherwise, the requirements for the doors are perfectly ordinary. A relatively wide door is convenient when moving equipment in and out. A screen-door closer is very useful to help prevent chicks from escaping. Using a door with a window in it makes it easier to check on the chicks without going inside.

Put a lock on the door, and use it. This is mostly to keep unsupervised children out of the brooder house. Even if you can trust all the kids who live in the neighborhood, you can't trust all their friends and relatives. Kids who have no experience with livestock can be awfully hard on baby chicks.

If you can hang a regular door, an ordinary entrance lockset is best. Otherwise, use a padlock.

14.8. Windows

I don't use glass windows in my chicken houses, but windows provide a nice way to control the ventilation in the brooder house. Double-hung windows are especially good, since you can open the top sash and leave the bottom one closed, minimizing floor drafts. When the weather gets hot or the chicks don't need heat anymore, you can open both sashes. In parts of the country that are hotter than mine, it's traditional to remove the sashes altogether in the summertime, once the chicks are off heat.

Too much glass area is a nuisance in chicken houses. Because chickens don't like wide daily temperature swings, glass is a bad idea, since it lets in a lot of heat during the day and lets out a lot during the night. Both a blank wall and a hole in the wall compare favorably to glass.

14.9. Type of House

Small poultry houses are usually shed-roofed, since such houses are easier and cheaper to build than gable- or combination-roofed houses. Once the house gets more than about 20 feet deep, you have to start thinking in terms of roof trusses, and gable- or combination-roofed houses start coming into their own.

The traditional brooder house in the old days was the 10x12' colony brooder house, which was built on skids, had a wooden

Figure 2. A typical old-time colony brooder house, 10×12 feet. The door is on the far side.

floor, and had a capacity of up to 250 chicks for the first four weeks.

My own brooder houses are only eight feet square, made out of the cheapest possible materials. I have a terrible tendency to procrastinate, and if I can't finish a project quickly, it may not get finished at all. I once left a chicken house half-finished for several months, and afterwards resolved to come up with a design that's so simple that I can build it in a single day. These extremely simple henhouses lack such features as insulation and paint, but they have proven very reliable in service. I have 18 chicken houses altogether.

Though designed as henhouses, these are adequate as brooder houses if the chicken wire at the front is covered with tarps or plastic feed sacks, and if the gaps between the walls and the roof are mostly closed off.

Any kind of structure can be made into a brooder house. I have considered converting single-wide mobile homes (which one can pick up for next to nothing) into a brooder house. They already have insulation, plumbing, and electricity. I would probably want to add a layer of highly rot-resistant plywood to the floor

Figure 3. A gable-roofed colony house, 8x12 feet. For brooding, cloth would be stapled over all of the door and window openings, and peeled back gradually as the chicks grew.

Figure 4. My two brooder houses, which are really just rudely converted henhouses. The one of the right is buttoned up for little chicks with tarps and feed sacks, while the one on the left has its window uncovered for older chicks who are off heat. The trash can holds chick starter.

and also to the bottom 24 in. or so of wall, and the window space may be inadequate for the latter stages of brooding, but that's nothing I can't fix with a reciprocating saw.

Many localities are too gentrified for chicken houses as crude as mine, and the neighbors would be up in arms if anyone presumed to install one without a full complement of gingerbread, weather vanes, window boxes, and other indications that your chickens are creatures of wealth and refinement. This doesn't seem to be an issue in my neighborhood.

14.10. Practice Perches

Practice perches are small perches used to teach the chicks to roost early. Chicks are born with the instinct to huddle if they are cold or frightened, which leads to deaths due to suffocation. When the instinct to sleep on perches emerges, it seems to replace the instinct to huddle. Chicks that learn to perch early are less liable to losses due to huddling.

Perches also get them up off the floor, which keeps them cleaner, and increases the effective floor space, since it gives the brooder house an upper level, as it were.

Figure 4 shows a typical practice perch setup. The roosts are slanted down to floor level to encourage the chicks to use them before they can fly. The sides and bottoms are covered with chicken wire to keep the chicks from hiding underneath the perches. These are traditionally installed during the third week, though there's no reason not to install them before the chicks arrive, outside the draft guard, if that's more convenient.

A small brooder house may not have room for all the brooding equipment plus a set of practice perches. Mine don't. I put a row of more conventional perches at the back of the brooder house, about 24 in. off the floor. These are simple eight-foot lengths of 2x2 lumber. My chicks start roosting well at four or five weeks of age, which I find adequate.

The chicks will start by using the roosts by day and sleeping under the brooder at night. When they don't need the brooder heat anymore, they'll spend the night on the perches. This gives you an easy way of determining when the brooding period is over.

If you don't provide perches, the chicks will continue to sleep on the floor in the vicinity of the brooder, and there's no obvious indication that the heat is no longer necessary.

Modern broilers are generally not provided with perches. They can't fly well enough to make use of conventional high perches, and in the old days there were a lot of problems with crooked breastbones and breast blisters attributed to the perches. The former problem turned out to be nutritional. The latter one may still be a problem. I have often wondered if low perches (low enough for broilers to step onto without jumping) might be a good idea, but I haven't tried it.

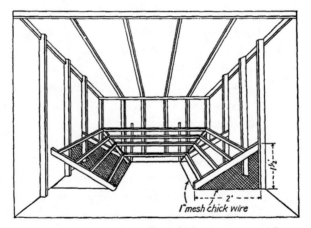

Figure 5. A practice perch setup that will get chicks roosting as early as possible. (Illinois Agr. Exp. Sta.)

Chapter 15. Health Problems and Predators

There are plenty of books that list every chicken disease known to man. I recommend Gail Damerow's *Chicken Health Handbook*. Most health problems, though, are so rare that you will probably never see them.

In this chapter I will cover the handful of problems that you will probably run into sooner or later.

15.1. Roost Mites

Roost mites are a tiny blood-sucking parasite. Like all mites, they are not insects but arthropods, though I don't suppose the distinction is of much practical use to poultrykeepers. Roost mites live in cracks and crevices in the chicken house or in the litter, and generally attack the chicks at night, though they aren't particular.

If present in large enough numbers, roost mites can kill chicks.

Detecting roost mites often starts when you have a crawly sensation up your arms or legs after visiting the chicken house. As far as I know, the mites aren't dangerous to humans, but they're disgusting. But they're dangerous to the chicks.

The mites themselves are quite tiny. They're gray in color before they feed, and red afterwards.

The shed skins of roost mites make a greyish, salt-and-pepper residue in places where they've been living. The roost mites themselves huddle in blood-red masses on the undersides of roosts, in between studs and siding, in the litter, and in other sheltered places.

Controlling roost mites is easy with insecticides. Both Sevin and Malathion are cheap, readily available, and do a good job.

For roost mites in the litter, powdered Sevin or Malathion are appropriate. Buy the stuff in one-pound shaker cans at the feed store. These will have directions for use in a poultry house.

If you're running an organic operation, you are restricted to "natural" insecticides on the approved list. One of these is pyrethrins, a naturally occurring nerve poison, which, like Sevin and Malathion, is a lot more toxic to bugs than to people or poultry. Buy it in dust form and follow the directions.

To control mites in the walls and roosts, you can use the liquid form of your favorite insecticide. I've also had good results with lime-sulfur spray. Oils will suffocate them. A 3% solution of miscible tree-spray oil is supposed to be very effective. Both lime-sulfur solution and tree-spray oil ought to be obtainable at any garden supply store.

You can also kill both mites and their eggs with sufficiently hot water. A hot-water pressure sprayer would come in handy here.

In general, you'll have to treat both the litter and the house with their respective bug killers, but you can treat just the house if you remove all the litter first. This is not practical if there are chicks in the house.

The eggs of the roost mites are typically not killed by insecticides. The ones I've mentioned all have low persistence, and will have decayed into harmlessness by the time the eggs hatch after about five days. So a second treatment 5-7 days after the first is a good idea.

Roost mites are spread by wild birds. If your brooder house excludes sparrows and such as well as it ought to, and you remember to keep the door closed at all times, even when there aren't any chicks inside, you may be able to avoid them. With my free range hens, who are constantly exposed to wild birds, roost mites are a fact of life, but I haven't seen any roost mites in the brooder houses for several years now.

15.2. Coccidiosis

Coccidiosis is an infection in the digestive system, caused by protozoan parasites called coccidia. There are a number of different species of coccidia that affect chickens. Coccidia are everywhere,

Figure 1. Chick with coccidiosis.

and severe outbreaks can occur in brand-new houses with brand-new equipment that has never seen a chicken.

Coccidiosis normally strikes chicks between three and seven weeks old. Survivors will become immune to the particular species of coccidia they were infected with. In some cases there is cross-immunity, so exposure to one species give resistance to another one.

Coccidiosis not only kills some of the infected birds, but permanently damages the digestive systems of survivors, stunting them.

Coccidiosis spreads through egg-like "oocysts" that come out in the chick's feces. These are harmless if eaten right away, but if they have time to develop ("sporulate"), they will infect a chick that eats them. Sporulation takes several days. If the infested manure is removed before then, the cycle of infection is broken.

As more and more chicks become infected, there is a population explosion of oocysts in the feces, and the chicks ingest larger and larger numbers of sporulated oocysts. They can ingest some without getting sick, but large numbers will overwhelm their defenses.

Coccidiosis can be prevented by minimizing the amount of feces that the chicks ingest. Keeping the feeders and waterers free of litter is a good start. If the feeders go empty, the chicks will forage in the litter more, increasing exposure. Feed spilled onto the floor will also increase exposure.

Brooding the chicks on wire floors instead of litter will cause the manure to fall through the floor, out of reach. This is very effective in preventing coccidiosis.

Though labor-intensive, removing all the litter from the brooder house every two or three days will usually work.

Another method of controlling coccidiosis is by making the physical, chemical, or ecological balance of the litter hostile to the oocysts:

* Built-up litter has anti-coccidial properties after about six months.
* Alkaline litter (produced by stirring in 10 pounds of hydrated lime per 100 square feet of floor area) seems to enhance the effects of built-up litter.
* Dry litter reduces coccidiosis.

Finally, medicated feed will prevent coccidiosis.

The symptoms of coccidiosis range from almost invisible to horrifyingly obvious. Typical symptoms involve lethargy and droopiness on the part of the chicks, loss of appetite, perhaps bloody droppings, and an increase in mortality. (Normally you don't really expect any mortality in chicks between three and seven weeks old, except with broilers, where the last couple of weeks in the period often have some growth-related losses.)

I recommend that beginners use medicated feed. For one thing, it takes six months for built-up litter to acquire anti-coccidial properties, which means there is a window of vulnerability that must be passed. Also, we don't want your first experiences with chicks to be marred by outbreaks of disease. Switch to non-medicated feed in your second season if you like.

I use medicated feed with my egg-type pullets, while Karen uses non-medicated feed with her pastured broilers. The big difference is that her broilers get moved out of the brooder house and onto pasture no later than 21 days of age, and often at 14 days of age. My chicks stay in the brooder house for 6-8 weeks. Her chicks just aren't old enough to have gotten sick with coccidiosis by the time they leave the brooder house, while mine are.

While both of us use built-up litter, I don't trust mine very much because I often fail to put my chicks on range at the proper time. The birds continue to grow, the brooder house becomes overcrowded, and the litter becomes wet. So I use medicated feed as a way of preventing my bad habits from harming my chicks.

15.3. Feather-Picking, Toe-Picking, and Cannibalism

Chicks sometimes peck at and even kill one another. The outbreaks I've seen start with chicks plucking out each others' tail feathers, which quickly escalates to their actually killing each other. Another form involves chicks pecking at each others' toes.

Cannibalism sometimes strikes out of the blue. It can happen to anyone. But it is particularly likely to happen if the chicks are overcrowded. Other factors that increase cannibalism are lack of feeder space or feed running out, high light levels, low-fiber diets, brooding on wire floors, and using highly cannibalistic strains of chicken.

Some strains of chicken are much less cannibalistic than others. For several years I made a point of ordering only the least-cannibalistic strains offered by my hatchery. I recommend this practice.

A friend who had a serious outbreak of cannibalism reported that reducing the light levels in the brooder house worked very well at curbing the outbreak. Nothing else worked at all. Red light does not promote cannibalism the way white light does. Old-time poultry manuals often recommended that window panes be sprayed with red lacquer if cannibalism was a problem.

Giving the birds more space, especially by moving them to range, will prevent or stop outbreaks.

My experience is that the first sign of cannibalism is the presence of tail feathers on the brooder-house floor. This usually precedes actual blood-letting by at least a few days, so keep your eyes open and be prepared to take steps.

The traditional advice for treating injured chicks is to coat the affected parts with pine tar and to release them back into the flock. I have never tried this.

Cannibalism can break out in chicks as young as three weeks, though I have never seen it in chicks younger than five or six weeks. Modern broilers seem to be non-cannibalistic. The same can't be said for some strains of modern layers.

15.4. Predators

15.4.1. Rats

I didn't have any problems with rats during my first five years of poultrykeeping, but I suddenly had lots of trouble after that. Rats found their way into one of my brooder houses and killed quite a few chicks before I realized they were there. Rats will kill chicks up to five or six weeks of age and drag them away, hiding the bodies in tunnels. If you don't count your chicks frequently (and I don't), you can lose a lot of them before you realize that they're aren't as many as there used to be.

My brooder houses were parked on a concrete slab, but the rats came through the walls below the level of the litter, and left no holes that were obvious from the inside. I set out rat poison around the houses and used galvanized steel flashing to armor the walls. After a couple of false starts, the rats were excluded and I have suffered no more losses.

There are a certain number of rats almost everywhere. Their numbers will skyrocket if you accidentally give them a food supply and safe places to nest. The traditional advice for controlling rats is to clear away all equipment, brush, weeds, and trash from near the brooder houses to deny them safe places to nest, to make it hard for them to get into the brooder houses, and to use poison bait, properly protected so it isn't eaten by any but the intended species. Information on the proper use of traps and bait is available from many sources; I won't repeat it here.

Poisoning is unpleasant, but having rats destroy a batch of baby chicks is worse.

Cats will help keep the rat population down. They tend not to tangle with full-grown rats, but they cut quite a swath through the younger ones. Some terriers are excellent ratters, I'm told, and will eliminate the on-farm population of adult rats.

15.4.2. Cats

I have lost a few chicks to cats, though not in recent years. Adult chickens can hold their own against cats. My hens will sometimes walk up to my barn cats' feed bowl and peck at the cats until they go away, then steal the cat food.

I think that my cats lost interest in baby chicks after encountering hens that had hatched a brood on their own. Being attacked by a protective mother hen may be the reason why they avoid chicks these days.

In general, you should exclude cats from the brooder house. I have had cats kill chicks up to six weeks of age.

15.4.3. Dogs, Foxes, and Coyotes

Dogs, foxes, and coyotes will happily kill every chick in the brooder house, but if you brood your chicks in confinement, they probably will never have the chance. Putting a screen-door closer on the brooder house door will help prevent the neighborhood dogs from getting even a fleeting opportunity of getting into the brooder house while your back is turned.

15.4.4. Raccoons

Raccoons are the burglars of the animal kingdom. They wear masks and everything. Raccoons are good at breaking into chicken houses. In fact, they can sometimes kill chickens even without breaking in. If there are chickens sleeping close to a window, they will reach through the chicken wire, grab the chicken, and tear it to pieces, pulling wings and legs through the wire. This has happened to me.

I have also heard stories of raccoons prying chicken wire loose from windows, but this has never happened to me.

To exclude raccoons, make sure that the doors of your brooder houses close securely, and that a raccoon-sized creature can't squeeze in. Any openings that a chick might conceivably sleep against should be covered with ½" or ¼" hardware cloth. 1" chicken wire is far too big for safety.

Raccoons are a big reason why chicks should be brooded in confinement.

15.4.5. *People*

I've found that children can be a big threat to baby chicks. Many children find chicks attractive and will play with them if they get the opportunity. Sometimes they play gently and sometimes they play rough. This can be hard on the chicks. A padlock on the brooder house is the best protection.

Inviting the neighborhood children to take a look at your baby chicks gives you the opportunity to talk to them about proper handling, but this won't rub off on kids who don't live in the neighborhood, but drop by sometimes to visit friends and relatives.

In the old days there used to be a lot of chicken thieves. Back then, everybody knew how to butcher a chicken. These days, most criminals would turn up their noses at such work. Still, there are some neighborhoods where the light-fingered element remembers its roots. Keeping the brooder house locked and close to your home will help prevent nasty surprises.

Chapter 16. Useful Tables

The following is all from Ewing's *Poultry Nutrition*, 3rd Edition, 1947, pp. 1321-1342. The values will be accurate for most breeds, but not for Cornish-cross broilers or bantams.

16.1. One Hundred Chicks Will Eat...

10 pounds of feed the first week.
20 pounds of feed the second week.
30 pounds of feed the third week.
40 pounds of feed the fourth week.

100 pounds the first 4 weeks.
360 pounds first 8 weeks.
765 pounds the first 12 weeks.
1,255 pounds the first 16 weeks.
1,825 pounds the first 20 weeks.
2,475 pounds the first 24 weeks.

16.2. One Hundred Chicks Will Drink...

1 gallon per day at 1 week.
2 gallons per day at 2-4 weeks.
3 gallons per day at 5-8 weeks.

16.3. Brooder Capacity

Here is how to figure how many chicks can be brooded in a conventional brooder with a canopy or hover: When brooding for a full brooding period (over six weeks), figure 14 square inches per chick, or 10 chicks per square foot of canopy. With a short brooding period (4-5 weeks), figure 7 square inches per chick, or 20 chicks per square foot.

Chapter 17. Recommended Reading

17.1. In-Print Poultry Books

The books in this list cover everything from backyard poultry-keeping to full-scale commercial operation.

Chicken Health Handbook, by Gail Damerow. Storey Publishing, 1994. This book tells you everything about chicken diseases, their identification, prevention, and cure.

Pastured Poultry Profits, by Joel Salatin. Chelsea Green Pub. Co., 1996. If you're thinking about raising meat chickens, even just for your own use, you should read this book. It gives step-by-step instructions for raising pastured broilers as a sustainable small-farm business. Designed for beginners, these methods really work. We've used them on our farm for years.

Storey's Guide to Raising Poultry: Breeds, Care, Health, by Leonard S. Mercea. Storey Publishing, 2000. Originally published as *Raising Poultry the Modern Way.* This book is the best overview of raising chickens, turkeys, ducks, and geese. Covers a lot of ground but necessarily skimps on detail. One of the few general poultry books still in print written by an Extension Poultry Specialist.

Storey's Guide to Raising Ducks, by Dave Holderread. Storey Publishing, 2000. Previously published as *Raising The Home Duck Flock.* Dave's waterfowl hatchery is just a few miles from my farm and we've been very happy with the waterfowl we've bought from him. His book is also excellent, going into detail on every aspect of duck raising. Highly recommended.

Commercial Chicken Production Manual and *Commercial Chicken Meat and Egg Production,* by Mack North and Donald Bell or Donald Bell and William Weaver. These are two successive editions of the same book, a massive reference to current commercial chicken practices. This book is ludicrously expensive (try to read a copy through inter-library loan), but it answers a lot of questions if you're raising chickens commercially, even on a small scale.

Commercial Poultry Nutrition, by Steven Leeson and John D. Summers. University Books, Ontario, 1997. An excellent reference book for anyone who wants to invent his own poultry rations and can afford the book's high price. The focus is on the creation of practical poultry feeds, not the esoterica of poultry digestion. Quite readable and accessible, given the subject matter.

Poultry Breeding and Genetics, R. D. Crawford, Editor. Elsevier Science, 1990. Another shockingly expensive book aimed at the well-heeled professional. This is the definitive reference to poultry genetics. A must for the die-hard poultry breeder. This book is a compendium of chapters written by different authors, and is not aimed at the interested layman, but it is thorough. Hutt's *Genetics of the Fowl* is a much better book, but Hutt is fifty years out of date, out of print, and hard to find.

17.2. Out-of-Print Poultry Books

Though many of these books have been out of print for more than fifty years, but they are still valuable references, especially to people with small flocks. Small flocks used to be the norm, and the first fifty years of research by poultry scientists focused on the needs of small flocks and small farms. Poultrykeeping on a small scale hasn't changed much over the years. Also, the poultry books of yesteryear were aimed at a more general audience, not highly trained specialists.

Books written after 1940 are usually quite accessible to a modern audience. Books written before 1900 usually spend most of their time on issues that are no longer relevant to us, because the understanding of diseases, breeding, and nutrition was so poor back then. Books between 1900 and 1940 fall somewhere in between.

To read out-of-print books for free, I use the inter-library loan service at my local library in the rare cases when I can't find the book at Oregon State University's Valley Library.

To buy out-of-print books, I rely on Abebooks (http://www.abebooks.com). Some of these books can be found at almost any time, but for others you'll want to set up an item in your "want list" so you'll get an e-mail when one becomes available. So far, this has always worked for me within a few months. Even the most obscure books hit the on-line booksellers once in a while.

Battery Brooding, Milton H. Arndt, Orange Judd Press, 1930. The definitive work on battery brooding. I've quoted sections in this book, but there's plenty more where that came from. Easy to find on used-book Web sites.

Successful Poultry Management, Morley A. Jull, McGraw-Hill, 1943. A practical guide to farm poultry, more accessible than most. Easy to find on used-book Web sites.

Practical Poultry Management, James E. Rice and Harold E. Botsford. Wiley, 1925-1956. This book had a lot of staying power and went through six editions. The last one is probably the best. It's also a good, practical guide to poultry, and is easy to find.

The Dollar Hen, Milo Hastings, Arcadia Press, 1909. While probably not to everyone's taste, this is my favorite poultry book. It covers free-range poultrykeeping and practical poultry farming in a concise, practical manner, and provided me with useful knowledge and inspiration over 90 years after its publication. Hastings became the poultry scientist at the Kansas Experiment Station, but was given no facilities for poultrykeeping experiments. So he decided to do a survey of how the poultry industry actually operated, from actual farm practices through wholesalers to retailers to the end user. This gave him a practical understanding of all aspects of the industry, which he imparts with vigor and acid humor. Hastings totally lacks the romanticism and impracticality of many poultry writers, and goes out of his way to kick them in the pants a few times. Hastings is also the only poultry scientist to write a classic work of science fiction (*City of Endless Night,* 1919). He later became active in the "physical culture" health-food and exercise movement of the 1920s. This book is hard to find.

Poultry Breeding and Management, James Dryden, Orange Judd Press, 1916. Dryden was the first person to prove that chickens could be bred successfully for higher egg production (earlier attempts by other scientists had failed). His experiments took place a few miles from my farm, on the campus of what was then Oregon Agricultural College (now Oregon State University). The book is really two books in one, with independent sections on breeding and management. While the breeding section is fascinating, it's the management section that makes the book worth seeking out. Like Hastings, Dryden was keenly aware of the realities of farm poultry flocks, and devotes several hundred pages to practical advice. At the time, most farm flocks were given free range, and this book is an invaluable reference for free-range and backyard poultrykeepers today. American farmers obviously thought so, too, since the book remained in print for over thirty years in spite of never being updated, and I have heard claims that it was the most popular poultry book ever written. This book is easy to find at low prices on used-book Web sites.

Turkey Management, Stanley Marsden and J. Holmes Martin, The Interstate, 1939. This is the book for raising turkeys, especially in small flocks or on free range. This book is crammed full of useful information and techniques. I've even applied some of their advice to my chicken flock with good results. This book is fairly easy to find on used-book web sites.

Poultry Breeding, A. L. Hagedoorn and Geoffrey Sykes. Crosby Lockwood, 1953. A clear, insightful, practical, and engaging book covering all aspects of practical poultry breeding. It covers both utility strains and show birds, and shows a deep understanding of the different needs of these two types. Hard to find. Hagedoorn's *Animal Breeding* is easier to find and almost as useful to the poultrykeeper.

Genetics of the Fowl. Hutt, 1947. The best book on poultry genetics, though long out of date. It complements Hagedoorn & Sykes' book; read them both and you're an instant expert in breeding and genetics. Modern books are more up to date but don't do anywhere near as good a job in presenting the fundamentals. Read these and you'll know more than someone than someone who has just read the modern equivalents. Unfortunately, *Genetics of the Fowl* is very hard to find and usually sells for $100 or more.

Feeding Poultry. G. F. Heuser. Wiley, 1946. An excellent handbook of poultry nutrition, with a lengthy chapter on green feed and the nutritional value of free range. This book complements the more modern *Commercial Poultry Nutrition* due to its understanding of small-farm conditions. Later editions of *Feeding Poultry* from the mid-1950s are probably best, as they add information about vitamin B-12 and medications that were unknown in 1946. This book is easy to find.

17.3. Small-Scale Farming Books

These books are of interest to most people with small farms. All are excellent.

The Encyclopedia of Country Living, by Carla Emery. Ninth Edition. Sasquatch Books, 1990. A monumental work with sections on just about everything, from livestock care to cooking, mixed with homespun philosophy. You can open it up at any point and start reading, and you'll be hooked. The poultry section (which I'm best qualified to comment on) is very well thought-out and practical.

You Can Farm: The Entrepreneur's Guide to Start and Succeed in a Farming Enterprise, by Joel Salatin. Polyface, 1998. This is Joel Salatin's guide for getting started in farming. It came out after we'd been on the farm for several years, and it matches up very well with our own experiences. We could have used it when we were starting out! Highly recommended for those who are even toying with the idea of farming.

Five Acres and Independence: A Handbook for Small Farm Management, by M. G. Kains. Dover Publishing. A classic guide to making a living from a very small farm. Kains emphasized the importance of direct marketing decades before it became an industry buzzword. He recommends a diversified farm with a succession of products throughout the season, with "anchor" products of greenhouse plants, berries, fruit, and eggs.

Chapter 18. For More Information

Please visit my Web site at http://www.plamondon.com. It has additional information about poultrykeeping, contains a section listing recommended hatcheries and suppliers, and gives information about my farm.

Printed in the United States
141999LV00003B/222/A